Sample Preparation Handbook for Transmission Electron Microscopy

Jeanne Ayache · Luc Beaunier
Jacqueline Boumendil · Gabrielle Ehret
Danièle Laub

Sample Preparation Handbook for Transmission Electron Microscopy

Methodology

Foreword by Ron Anderson

Jeanne Ayache
Institut Gustave Roussy
Unité mixte
CNRS-UMR8126-IGR
Laboratoire de Microscopie
Moléculaire et Cellulaire
39 rue Camille Desmoulin
94805 Villejuif CX
France
ayache@igr.fr

Luc Beaunier
Université Paris VI
UPR 15 CNRS
Boîte courrier 133
Labo. Interfaces et Systèmes
Electrochimiques
4 place Jussieu
75252 Paris CX 05
France
luc.beaunier@upmc.fr

Jacqueline Boumendil
Université Lyon I
Centre de Microscopie
Electronique
Appliquée à la Biologie et à la Géologie
43 bd. du 11 Novembre 1918
69622 Villeurbanne CX
France
jb.boumendil@gmail.com

Gabrielle Ehret
Université Strasbourg
CNRS-UMR 7504
Inst. Physique et Chimie des Matériaux
22 rue du Loess
67034 Strasbourg CX 2
France
jcehret@evc.net

Danièle Laub
Ecole Polytechnique Fédérale
de Lausanne
Faculté des Sciences de Base
Centre Interdisciplinaire de
Microscopie Electronique
039 Station 12
1015 Lausanne
Bâtiment MXC
Switzerland
daniele.laub@epfl.ch

ISBN 978-1-4899-8697-9 ISBN 978-0-387-98182-6 (eBook)
DOI 10.1007/978-0-387-98182-6
Springer New York Dordrecht Heidelberg London

Cover illustration: Conception: Dan Perez
TEM image of freezing defects in a frozen thin film, showing clusters of segregated crystals along the holes of the carbon membrane. (Baconnais S., CNRS-UMR8126, Villejuif, FR).

Printed on acid-free paper

Springer is part of Springer Science+Business Media (www.springer.com)

The gift that microscopy brings us, beyond the beauty of the images, is that it gives us access to the Art of Matter and brings us to the heart of the mechanisms, "from the structure of inert matter to the complexity of the living."

There, where the infinitely small and the infinitely large come together...

It is a lesson of life and humility.

Jeanne Ayache

Foreword

Successful transmission electron microscopy in all of its manifestations depends on the quality of the specimens examined. Biological specimen preparation protocols have usually been more rigorous and time consuming than those in the physical sciences. For this reason, there has been a wealth of scientific literature detailing specific preparation steps and numerous excellent books on the preparation of biological thin specimens. This does not mean to imply that physical science specimen preparation is trivial. For the most part, most physical science thin specimen preparation protocols can be executed in a matter of a few hours using straightforward steps. Over the years, there has been a steady stream of papers written on various aspects of preparing thin specimens from bulk materials. However, aside from several seminal textbooks and a series of book compilations produced by the Material Research Society in the 1990s, no recent comprehensive books on thin specimen preparation have appeared until this present work, first in French and now in English.

Everyone knows that the data needed to solve a problem quickly are more important than ever. A modern TEM laboratory with supporting SEMs, light microscopes, analytical spectrometers, computers, and specimen preparation equipment is an investment of several million US dollars. Fifty years ago, electropolishing, chemical polishing, and replication methods were the principal specimen preparation methods. Ion milling, tripod polishing, and focused ion beam (FIB) tool methods were yet to be introduced. Today, a modern ion milling tool can cost tens of thousands of dollars and a fully outfitted FIB tool can easily cost a million dollars. With investments of this magnitude – made necessary by the demands placed on modern TEM analysis – it is paramount that the staff preparing TEM specimens have all of the training and resources possible to carry out their duties. This is where the book in your hands comes in!

But thin specimen preparation is more than just laboratory hardware and excellent protocols for thinning a specimen to electron transparency. Successful thin specimen preparation also requires knowledge of the information required from the TEM analysis. The question determines the method! For example, there may be several methods that could be used to produce specimens of the same material, but without a clear idea of the information required, even successfully thinned specimens may be only marginally useful. Thus, considerable thought should go into understanding the problem. In some cases, information from light microscopy, SEM, powder X-ray diffraction, and a trip to the library (or at least to the Internet) will solve the problem without even making a TEM thin specimen. In other cases, such information will be helpful not only in the TEM analysis itself but also in preparing appropriate thin specimens for such analysis. Unlike analytical methods that routinely deal with completely unknown specimens, say powder XRD, successful TEM requires the analyst to bring considerable knowledge to the microscope – and even to the specimen preparation! It is more important to bring knowledge to the specimen prep. Wrong prep, and the scope time is useless. Thus, we should set some realistic goals for our thin specimens and bring as much intelligence to the table as possible. Here are three goals to consider no matter what material is to be thinned:

Goal 1: *To produce an electron transparent specimen representative of the bulk material in both structure and composition.* To meet this requirement, the researcher must have a comprehensive knowledge of the structure of the material system to be studied. It might be possible to produce an electron transparent specimen by beating your material with a hammer and collecting the thinnest shards for observation. However, it is likely that the resulting specimen would not have any relationship to the structure of the material before it was so "processed." The writer is certain that there are researchers working with silicon semiconductor specimens who think that the microstructure of single crystal Si contains numerous "salt and pepper" small defects that are actually ion milling artifacts. A good rule of thumb to follow is to prepare TEM specimens by more than one method if possible. Comparing ion-milled Si with chemically polished Si thin sections will immediately establish the true microstructure of Si, for example. The well-prepared analyst should know that as-grown single-crystal Si should be featureless and that an Al–Cu alloy will contain a family of precipitates as a function of the specimen's thermal history. Facts like these on any system to be studied may be found in the literature or learned in discussion with colleagues. This handbook provides clear instruction on the many specimen preparation methods by which it should be possible to produce alternative studies of a given material – with the advantages, disadvantages, and artifact risks of each – so that an analyst can deduce the true microstructure of their specimen.

Goal 2: *To provide easy access to the required specimen information.* This would not be a problem if the specimen preparation protocol always yielded "ideal" specimens. What is an "ideal" specimen? First, the transmission electron microscopy specimen must be thin. How thin? Optimum thickness varies with the microscopy application and the information desired. The optimum thickness for dislocation density measurements may be 100 nm or greater, but the optimum thickness for electron energy loss spectrometry measurements is often less than 10 nm. The ideal specimen should maintain an ideal thickness over a large area. Second, an ideal specimen should be flat, strong, homogenous, and stable under the electron beam for hours and in the laboratory environment for years. Finally, an ideal specimen should be clean, conducting, and non-magnetic. The reader may well conclude that there is no such thing as an ideal specimen. Compromises have to be made. Perhaps no single specimen preparation method is perfect. Given a thin film alloy containing precipitates, for example, electropolishing might thin the alloy matrix but leave the precipitates too thick to analyze, whereas, ion milling might thin the precipitates but induce objectionable artifacts in the film matrix. Specimen preparation may also be limited by external factors. In the example just given, a focused ion beam (FIB) tool could prepare a satisfactory thin specimen exhibiting both the precipitates and the matrix. However, such a tool can be very expensive, and the analyst's laboratory may not have access to one. Thus, less-expensive methods must be found. Expertise in as many thin specimen preparation protocols as possible is a great advantage in any laboratory, hence the utility of the present handbook.

Goal 3: *To produce a thin specimen that enables the microstructure of the material to be accurately studied and convincingly illustrated in reports and peer-reviewed publications.* The end goal of thin specimen preparation is the production of new knowledge displayed as micrographs in publications. Correct, artifact-free exposition of the specimen microstructure is all that matters in the final analysis and will probably be the only thing recognized by the scientific community. That community, and the analyst's management, really will not care which or how many preparation protocols are employed. It is the artistic skill and the knowledge of the specimen preparer that counts, hence the value of the present handbook.

This book provides the novice with a grounding in the major specimen preparation methods in use today, assessing their merits, and identifying those modalities that are most likely to yield success. Experienced specimen preparers can use these protocols to find alternative ways to prepare their standard specimens. In addition, new requirements may become necessary, such as high-spatial resolution in the prepared thin specimen itself, where the locations of specific predetermined sites are required to be within 100 nm. Moreover, now it is often required to prepare thin specimens in much shorter times than a decade ago.

For the most part, this handbook serves the physical science community. However, there has been a trend in recent years for performing materials science analysis in biological laboratories – especially with the increase in work on biomaterials and biomimetics. So what do biologists do with materials samples? Where do they turn for specimen preparation help? I am suggesting that this book and web site are the place.

The authors have chosen a unique format for publishing their work. They originally considered a book in two volumes with a companion CD. This static approach, where readers would wait between editions to learn new content, was abandoned in favor of a handbook with a companion dynamic web site, where the content can be updated as soon as new material appears. As fully explained in this handbook, the researcher is provided with web-based guides containing both a database of materials and an "automated route" to lead to the most appropriate specimen preparation technique based on sample properties and the choice of microscopy technique. The web content is extended via links to international microscopy centers and databases. The short files on the web site are augmented by the extensive treatment each topic receives in the book. You, the reader, can be part of this novel pedagogical approach;

there are facilities whereby you may add updates and new content to the web site as you develop them. Manufacturers making specimen preparation tools and supplies may also contribute to the project. This remarkable work will remain current and provide continually increasing value to the specimen preparation community.

Executive Editor, *Microscopy Today* Ron Anderson
IBM Analytical Laboratory, East Fishkill, New York (retired)
Fellow of the Microscopy Society of America and Past President
Largo, Florida
September 2009

Preface to the English Edition

It was a real adventure for our special club of five. Jeanne Ayache selected four collaborators for our supposed expertise in different areas of sample preparation and our belief that we really owed this "little job" the preparation of a guide to sample preparation, to our young and new colleagues. It is always attractive to share the experience of a career, and anyway the project (we thought!) could be completed within a year. Five microscopy specialists, each working in a different discipline and having a long-standing practice of teaching courses in this field, constituted a one-of-a-kind team.

With 5-times-20 years of experience, which, as they say in finance, comes to 100 years in accumulated surplus, our collaboration could not be reduced to a little 200-page manual. As the meetings went by, the program took shape, not without pains, resulting in a web site (in French and English) and the volume that we offer you today.

The first difficulty of this project was the language. Although we all speak French, we very quickly came up against our personal jargon: the "dialects" of a lab or of a scientific community (physicist, biologist, chemist, etc.). The richness of the French language is such that translations from French into English are different from one field to another, and habits are thrown in. For example, physicists talk about microstructures down to the scale of the nanometer, while biologists talk about ultrastructures and often stop at the scale of a tenth of a micron. Biologists who practically perform nothing but ultramicrotomy talk of "cuts," while physicists prepare "thin slices," even when they are making cuts! It almost felt like being in the tower of Babel. In short, we first had to create a glossary with a definition that provides exactly the meaning ascribed to the word used. This was a task that called for many debates and all our energy during long meetings. Once this primordial step in any interdisciplinary or cross-disciplinary undertaking was completed, everyone drafted the sections on techniques they practice frequently and know well.

The second unique aspect of the project was the collective reading of the various techniques, always with an "uninitiated" member in the group who knew nothing about the field being introduced. How should you explain an electrochemical manipulation to a biologist and an immunolabeling to a metallurgist, for example? The result is a selection of expressions accessible to all, including the non-specialist, at the expense of a super-precise aspect, of course. The techniques that we present here are written so that they may be understood by those who have never practiced them. We not only give you the outlines that make it possible to understand their implementation, their limits, and their artifacts, but also often present the details that enable their success. However, it seems difficult, for some techniques at least, to head to the workbench for an initial test, regardless of how complete the description is. Implementing a technique is not an “intellectual” task, but rather a technical task that can only be well learned in a practical training course. Our descriptions must enable you to choose which training course will be best adapted to the problem presented on the given material. Thanks to our shared experiences, we have listed the limits and imperfections of the techniques discussed for many types of materials. However, we do not claim to present all the variations and adaptations of techniques that may have been developed here or there with success.

Everyone knows the techniques most commonly used in their field, but do they know the ones used in other disciplines? Curiously, we realize that the process leading to the selection of the technique is the same in all disciplines: knowledge of one’s material, the methods of action of the techniques considered, and the requirements of the mode of observation planned. We also realize that a technique considered classic in one discipline may be poorly known in other scientific areas. Ultramicrotomy is probably the best example of a technique that had been bringing joy to biologists for the past 50 years before materials researchers became aware of its strengths as well as its limitations. By knowing the actions coming into play in each type of technique, we invite you to think about what is going on during preparation. This will enable us to predict whether or not our material will be damaged by preparation. We thus train our critical minds by improving the recognition of artifacts and refining the interpretation of our results. Technique is just like cooking, but scientifically reasoned cooking has a much greater chance of being effective and reproducible.

Today there are still too few interdisciplinary bridges due to a lack of relationships, communication difficulties, and/or hyper-specialization. But these bridges are essential to resolve the problems of materials that grow more and more complex and often involve mixed and composite materials. This work is aimed at the latest generation of microscopists, the researchers in emerging disciplines who need to characterize their new materials, and industrial researchers who are often confronted with never-before-seen problems that are sometimes far removed from their base training. In this compilation, they will find the ideas that are indispensable to understanding their problems and the means for solving them. This work might also be of great service to those who make it their calling to be open to all, such as technical platforms and joint imaging and analysis centers.

Yes, this was an adventure that carried us through 5 years of work in spite of ourselves. From being highly professional, our meetings also became very friendly, with bitter and heated debates to be sure, but always in the spirit of serving science rather than some personal flattery. Oh, how many things we learned in the course of those 5 years! First, in the disciplines that we were not familiar with, in the strictness of expression striving for a more universal language, and last, in the art of using all of the resources of a computer, including those for maintaining long-distance relationships between the various partners. Many times we had to go back to the drawing board, or to colleagues, to confirm an idea or illustrate a proposal. We would like to thank them wholeheartedly for their diligent and effective assistance. It was a lovely undertaking and a truly shared one, with each bringing their skills to the service of the common cause. It was a marvelous human adventure that will leave its mark on our professional relationships.

We would like to thank our various supervisors for agreeing to give us the time to do this and for the two retirees, thanks goes to their families for understanding the worthiness of this commitment. We also thank those who helped us technically speaking, including Michel Charles and the CNRS-Formation department in the creation of the web site preparation guide, Frédéric Lebiet for setting up the web site, Avigaël Perez for creating the diagrams, Bernard Lang for translating the web site sheets into English, Aurelien Supot and Michael Healey from Atenao Company for the translation of the French version of the books into English, and Joseph McKeown (Arizona State University, Tempe, USA) for the review of the final English manuscript.

Our gratitude most especially goes out to our colleagues of the LM2C laboratory of CNRS UMR 8126 at IGR and the CIME of EPFL of Lausanne, for their moral support, their help, and their precious advice on the creation of this collective work. We would like to thank those who supported us morally and financially in our undertaking: CNRS-Formation and the French Microscopy Society. Last, Gérard Lelièvre, Director of the MRCT of the CNRS, deserves special recognition. He supported us very early on in our approach and gave us the material means for this creation. We owe the publication of this book to him.

Villejuif, France — Jeanne Ayache
Paris, France — Luc Beaunier
Villeurbanne, France — Jacqueline Boumendil
Strasbourg, France — Gabrielle Ehret
Lausanne, Switzerland — Danièle Laub
October 2009

About the Authors

Jeanne Ayache CNRS researcher in materials science and biology, Molecular and Cellular Microscopy Laboratory, CNRS-UMR8126-IGR Mixed Research Unit, Institut Gustave Roussy, Villejuif, France.

Jeanne Ayache is a CNRS physicist and microscopist researcher. Since she joined the CNRS in 1977, her research activities have been focused on studying the structure of materials belonging to the interdisciplinary fields of materials and earth sciences. She especially studied the structure of natural and industrial carbon-based nanomaterials, superconducting ceramics, oxide-based thin film, and heterostructures, down to the atomic or molecular scale. She is now working in the life science research field, at the Cancer Institute Gustave Roussy UMR 8126 of CNRS in Villejuif, France, where she is developing the aspects of electron microscopy in cell biology.

Luc Beaunier CNRS researcher in physics, Electrochemical Interfaces and Systems Laboratory, CNRS Exclusive Research Unit UPR15, Jussieu, Université Pierre et Marie Curie, Paris, France.

Luc Beaunier is a CNRS researcher in physics in the Electrochemical Interfaces and Systems Laboratory at the Université Pierre et Marie Curie, Paris, France. His research activities in the physical metallurgy fields are related to corrosion phenomena induced by chemical and physical defects in metals. His last research interest is surface-alloyed metals by light energy laser treatment. All these materials are characterized by electron microscopy and spectrometry analysis (TEM, SEM-FEG, EDS, PEELS).

Jacqueline Boumendil Research engineer in biology and microscopist at the Université Lyon l, technical director of CMEABG, the Center for Applied Electronic Microscopy in Biology and Geology at the Université Claude Bernard-Lyon 1, Villeurbanne, France (Retired).

Jacqueline Boumendil was technical director of the Center for Applied Electronic Microscopy in Biology and Geology CMEABG at the Université Claude Bernard-Lyon, Villeurbanne, France, and is now retired. The 37 years she spent in this center led her to study many normal and pathological biological samples, as well as structure of new polymeric materials. She has set up training in electron microscopy sample preparation techniques that she taught for over 20 years. She has been in charge of the development of these techniques and particularly the cryotechniques.

Gabrielle Ehret CNRS engineer in mineralogy and materials physics and microscopist, technical director of the Microscopy Department of the Mineralogy and Crystallography Laboratory, subsequently technical director of the Institute for Materials Physics and Chemistry, Strasbourg, France (Retired).

Gabrielle Ehret was technical director in transmission electron microscopy at the Laboratoire de Minéralogie et Cristallographie, then at the Institute for Materials Physics and Chemistry, Strasbourg, France, and is now retired. Since she joined the CNRS in 1970, her specialty has been the study of minerals, catalytic samples, and nano-carbon specimens. She was in charge of the transmission electron microscope training and teaching for the new TEM users and student research support.

Danièle Laub Director of microscopy sample-preparation at the Lausanne Federal Polytechnical School (EPFL), Department of Basic Sciences, the CIME, Interdisciplinary Electron Microscopy Center, Lausanne, Switzerland.

Daniéle Laub is technical director of microscopy sample preparation at the Lausanne Federal Polytechnical School (EPFL), Lausanne, Switzerland. Since she joined the CIME (Centre Interdisciplinaire de Microscopie Electronique) in 1988, she has been in charge of the development of sample preparation techniques for different types of materials (polymer, metal, semiconductors, ceramics, catalyst, etc.). She is responsible for sample preparation techniques training and teaching to new TEM and SEM users.

Contents

Abbreviations

In order to simplify reading, abbreviations are used throughout this work for analytical techniques. These abbreviations are listed below:

ADF	annular dark field: annular dark field imaging mode in TEM or STEM microscopes
AFM	atomic force microscopy
BF	bright field imaging mode in a multi-beam diffraction condition
CBED	convergent-beam electron diffraction
CTEM	conventional transmission electron microscopy
Dedicated STEM	dedicated scanning transmission electron microscope
DF	dark field imaging mode in a Bragg diffraction condition
EBIC	electron beam-induced current
EBSD	electron backscattered diffraction
EDS	electron energy dispersive spectrometry
EELS	electron energy loss spectrometry
EFTEM	energy-filtered transmission electron microscopy
ELNES	energy loss near-edge structure
EXELFS	extended energy loss fine structure spectrometry
FEG	field emission gun
HAADF	high angle annular dark field performed in STEM, is a Z-contrast atomic level chemical imaging mode
HRTEM	high-resolution transmission electron microscopy
LACBED	large angle convergent-beam electron diffraction
PEELS	parallel electron energy loss spectrometry
SEM	scanning electron microscope
STEM	scanning transmission electron microscope: SEM accessory in a transmission electron microscope
TEM	transmission electron microscope
WDS	wavelength dispersive spectrometry
Weak Beam	dark field imaging mode in a condition of weak-beam diffraction

Chapter 1
Methodology: General Introduction

This book is aimed at the entire scientific community (solid state physics, chemistry, earth sciences, and live sciences), to those who use transmission electron microscopy (TEM) to analyze structure in relation to the properties and specific functions of materials. This work is essentially dedicated to the recommended methodology for beginning the preparation of a sample for the TEM. In particular it stresses the approach to take in selecting the best technique by taking into account the material problem presented, the type of material, its structure, and its properties. It proposes the tools for the most appropriate preparation of samples for observing the true structure of the material. In this work, you will find general information on the classification of different types of materials, their physical properties, and their microstructures. The presentation of the different types of analysis and observation methods used in microscopy is a reminder to everyone about the possibilities presented by TEM microscopy. The analysis of the physical and chemical mechanisms involved in the various types of preparation techniques makes it possible to better understand the artifacts that may be left behind by them. The illustrations of the artifacts observed under the TEM created by the different preparation techniques will enable beginners to easily identify them. The comparison of results and analyses obtained from one single material, prepared using different techniques, will guide the user's selections leading to the final decision. Finally, we propose the combination of several techniques in order to solve complex preparation problems and obtain thin slices that can be analyzed under the TEM.

Part of this work is gathered together in the form of a methodological guide on the web site http://temsamprep.in2p3.fr. This interactive web site uses theoretical data to help the user directly in making their choice of technique, using detailed decision-making criteria for each technique. Using an interactive guide, the user can, starting with data on the physical properties of their material and the analyses to be conducted, then figure out which preparation technique or techniques are best suited. The gradual determination is made based on the information gathered: limitations, advantages, drawbacks, and artifacts induced by each technique. The interactive web site and this work complement each other, with the book providing a great deal of additional information on both techniques and their processes of use.

You will find in detail the approach and the underlying material approach to the methodological guide on the site, with all of the different theoretical supplements

J. Ayache et al., *Sample Preparation Handbook for Transmission Electron Microscopy*,
DOI 10.1007/978-0-387-98182-6_1, © Springer Science+Business Media, LLC 2010

on the mechanisms involved during the use of the various techniques. One chapter is reserved especially for the various preparation artifacts and another for illustrating the complimentary nature of certain preparation techniques.

Chapter 2 presents the microstructure of materials and their physical and chemical properties in connection with the general problems presented by studying materials.

Chapter 3 presents the principles behind different structural, chemical, or spectroscopic imaging methods, diffraction methods, and lastly chemical and spectroscopic analysis methods.

Chapter 4 presents materials issues and the different types of TEM and TEM/STEM analyses. It suggests an approach for tackling the study of a material, whether that study be morphological, structural, chemical, or spectroscopic, by trying to identify what scales of analysis are relevant to the problem presented.

Chapter 5 brings together all the physical and chemical mechanisms involved during the preparation techniques and, in particular, the principles behind the mechanical, chemical, ionic actions associated with the techniques these actions are involved in. This chapter also explains in detail the action leading to a change in the state of materials containing an aqueous phase and the actions leading to a change in the properties of a material. It also recalls the processes leading to a physical or chemical deposit, applied to the corresponding techniques.

Chapter 6 brings together the different types of artifacts created during the sample preparation steps as well as those artifacts formed under the effect of the electronic beam during observation.

Chapter 7 brings together all of the criteria that make it possible to select the most appropriate preparation technique in order to respond to a given material problem, based on the TEM analyses that you wish to conduct.

Chapter 8 presents comparisons among several preparation techniques using the same material in materials science and biology, illustrating the complimentarity of techniques and the importance of using several techniques.

Chapter 2
Introduction to Materials

1 Introduction

1.1 Origin of Materials

Natural materials such as organic matter, mineral matter, and living matter, along with artificial materials produced industrially, make up all of the materials found on the Earth. They all have a chemical composition and particular structure that give them specific properties or functions in relation to their surroundings or their formation conditions.

Natural materials are formed in a particular environment, under the diverse conditions seen in nature. These materials can be studied either in their original state or after being modified.

An artificial material is a compound manufactured by synthesis under known conditions that are selected to give it specific properties related to its field of application. Metal alloys, ceramics, and polymers are some simple artificial materials. New materials are often made of complex structures composed of mixed or composite materials.

1.2 Evolution of Materials

As an enabler of technology, materials research has a wide range of applications in the physical and life sciences and in our daily lives. Furthermore, the domains of physics and biology are now growing closer together as physics generalizes the principles and models presented by biological processes. For example, biological materials are being integrated into molecular electronics applications, although the complexity of new physical materials has not yet reached that of materials in the living world.

The study of structure and properties requires material's preparation that enables observation and characterization of the material's state at the moment of sampling. It may be difficult to keep biological materials in their original state because they are very often hydrated and must systematically be physically or chemically stabilized in order to survive observation in the transmission electron microscopy (TEM).

J. Ayache et al., *Sample Preparation Handbook for Transmission Electron Microscopy*,
DOI 10.1007/978-0-387-98182-6_2, © Springer Science+Business Media, LLC 2010

Regardless of the material's complexity, properties, or functions, it is necessary to master all steps of the preparation techniques for the TEM. It is equally important to know the limits of each technique, its drawbacks, and especially the artifacts that it may induce in order to be sure that the analysis reveals the true nature of the material.

1.3 General Problems Presented by Microstructure Investigations

When studying a material, the microscopist is confronted by the relationships between its physical, chemical, thermal, and dynamic histories. The conditions the material was subjected to will dictate its particular microstructure formation at different scales, and thus its physical, chemical, and/or biological properties.

Regardless of the material type, three main parameters can be presented in the form of a triangular diagram. Figure 2.1 shows these parameters: (i) microstructure, (ii) growth related to its surroundings, and (iii) properties, which are interdependent. If just one of these parameters changes, then the other two are disrupted, sometimes

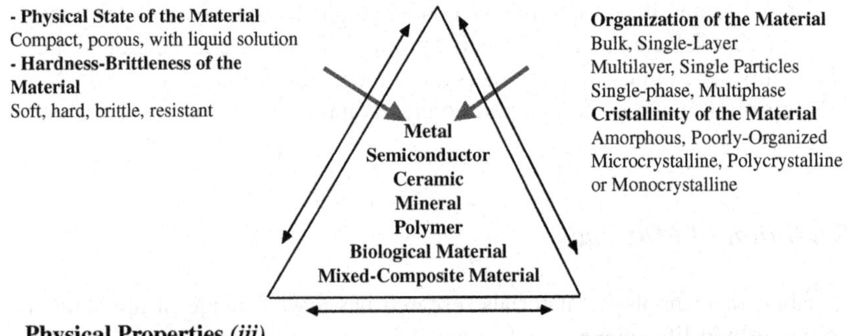

Fig. 2.1 Schematic representation of the problems to be resolved in materials science and biology when studying the microstructure of a material and the relationship between (i) its microstructure, (ii) the history of the material, and (iii) its properties

irreversibly. The challenge in developing new materials is to master all of the parameters of this system in order to reproduce the properties or functions needed for a specific application. In biology, the goal is to know all of the parameters of the living matter.

Diverse materials result from the natural evolution of a rock, mineral, organic material, or biological material or from the synthetic process for man-made materials. In addition, the mechanisms of growth or formation are different depending on whether materials are found in the solid state or liquid state or in intermediary solid–liquid states. Depending on the conditions of temperature, pressure, chemical gradient, kinetics of diffusion (atomic, ionic, or molecular diffusion), and the dynamics of the system, microstructures can be very diverse in materials science and biology.

A material's microstructure contains its structural organization on different scales. It also contains data on the chemical and structural phase distribution, the nature and concentration of defects formed, and the type of chemical bonds present in the material. All of these structures give a particular material its properties and functions. To understand their relationships, certain scales of observation are pertinent.

A given material type will be used for a particular function.

The interesting properties of a material are those that correspond to the physical and structural characteristics of the material or blend of materials.

Metals are used for their electrical conduction and their individual mechanical properties. Semiconductors are mainly used for applications in electronics due to their electronic structure. Ceramics, because of their very high fusion points, low density, the nature of covalent and/or ionic chemical bonds, and mechanical properties, will have a widely varying range of applications, from the manufacture of aerospace materials to electronic components. Ceramics are also often combined with other materials. Polymers are used in a large number of different fields from industrial materials to biomaterials.

In physics, interest will lie in "systems in equilibrium" when analyzing atomic structures. Indeed, in order to be sure of the material structure, it must be stable; in other words, it must have reached its state of equilibrium. Minerals are an obvious example. One may also need to study the dynamics of the system, which can be done artificially before observation or in situ in the microscope (see Chapter 4). Nevertheless, among the newer materials, multilayered nanostructured materials are far from being in thermodynamic equilibrium in their applied state.

Multilayer materials are composed of layers of different types of materials and have a 2D geometry that gives them very special properties tied to the proximity of the interfaces between the layers. Physical properties such as superconductivity, giant magnetoresistance, and ferromagnetism all correspond to the interaction mechanisms at the atomic scale. These mechanisms are associated with charge transfer mechanisms, magnetic induction, and electron spin exchange, and therefore deal with electronic structure. These materials have particular properties, often corresponding to different oxides that have variable oxygen concentrations or atoms that may have multiple charges. The combination of different types of materials having different structures and chemistry, as well as the proximity of interfaces, increases the interactions and eventually can result in new emerging properties. An atom's electronic structure or chemical bond is highly sensitive to the effects of proximity found at the interfaces between layers. Any change in the crystal structure near the interfaces will affect the atomic and electronic structure of the interface and will therefore affect the material's properties. Thus, mastering the particular physical properties of an application imposes a constraint on the technological quality of the development of these systems on the atomic scale.

The major difference between physical materials and biological materials is that living systems are dynamic. For example, the cell continuously adapts to its environment. To make this adaptation, the living organism (and its base unit, the cell), which is not an isolated system from a thermodynamic point of view, constantly gives off heat to its surroundings. Furthermore, biological materials are in liquid solution, which introduces a major difference from the viewpoint of the kinetics of reactions and interactions. In biology, there are several orders of magnitude within kinetics, from molecular movements to rates of reactions and therefore rates of functional transformations, which will involve ionic and electronic transfers. These times range from the femtosecond (10^{-15} s) (elemental chemical reaction) to the picosecond (10^{-12} s) (rotation of a water molecule), the nanosecond (10^{-9} s) (vibration of DNA base pairs), the microsecond (10^{-6} s) (molecular movement in DNA), the millisecond (10^{-3} s) (transcription and replication of a DNA base pair), and lastly to the second (1 s) (heart rate). These different timescales will result in continuous variations. Microstructure is never stable; analyzing microstructure in biology requires halting all movements and reactions and only allows one to see the structure at a given moment. In fact, what is important to know in biology is how the biological material maintains this structure over time and how the identifiable structural sites function.

The properties of biological materials are a result of constant changes in chemical equilibria through oxidation–reduction reactions; enzymatic reactions; ionic transport (very rapid diffusion in solution) through walls, membranes, or ion channels; and polymerization and depolymerization (e.g., what occurs continuously during cell division). Unlike materials common to materials science (which are unlike new nanomaterials that have multiple properties), biological materials always have multiple functional properties for a single microstructure. For example, an amino acid may be modified and involved in a metabolic or catabolic process at different times through the involvement of enzymatic systems or oxidation–reduction reaction systems. These mechanisms can occur in the same structural sites or in different sites, depending on the case.

2 Classification of Materials and Properties

2.1 Types of Chemical Bonds: Atomic and Molecular

The atoms and molecules comprising minerals and living matter are bound by six types of bonds with different intensities and properties. Examples include metallic bonds, covalent bonds, ionic bonds, and weak bonds. Among the weak bonds, there is a distinction between polar bonds or hydrophilic bonds (hydrogen bonds and van der Waals bonds) and nonpolar or hydrophobic bonds. From these properties will come the spatial form of the associated atoms and the molecules and then, at a larger scale, of the crystal, and finally of the organism as a whole.

Metallic bonds are formed by the sharing of electrons in the outer layer of the atom in an electron cloud, where they are free and delocalized. This free-electron gas ensures the cohesion of the remaining cations and enables electrical conduction in metals and alloys.

Covalent bonds are formed by the sharing of pairs of valence electrons in order to fill the outer electron shells of each atom. They are very strong bonds that are found in non-metals such as semiconductors, certain ceramics, polymers, and biological materials.

Ionic bonds are formed by the transfer of an electron from one atom to the other. They are strong bonds that appear, for example, between a metal atom that has released an electron and a non-metal atom that has captured the free electron. After bonding, both atoms become charged. These bonds are found in minerals, ceramics, biological materials, and certain polymers (ionomers).

Weak polar bonds are electrostatic and correspond to simple attractions between dipoles in compounds or molecules with inhomogeneous or polarizable charges. They act over long distances but with less intensity than strong bonds. Among them, for example, are *van der Waals bonds* between molecules and *hydrogen bonds* between water molecules in liquid water and ice. These bonds are found in all biological materials, certain hydrated minerals, polymers, and some mixed–composite materials.

Weak nonpolar bonds or hydrophobic bonds are formed by repulsion. In a polar liquid, the molecules try to establish a maximum number of bonds between each other. If nonpolar molecules are added to the solution, their presence disrupts the formation of this network of bonds, and they will be rejected. Uniquely nonpolar molecules are rare in nature and for the most part are found in hydrocarbons. Fatty acids are amphiphilic molecules, containing a polar end and a nonpolar end. These molecules will then form complex structures, with the polar end on the outside in contact with the water and the nonpolar end on the inside, completely isolated from the water. Depending on the nature of the molecule, these structures will either be small globules called micelles or be membranes. These bonds are found in all biological materials.

Among all materials, only biological materials, certain synthetic polymers, and certain mixed–composite materials or biomaterials have three types of strong, ionic, and weak chemical bonds coexisting together. The simultaneous presence of these three types of bonds gives the material a particular sensitivity to the effects of radiation and to thermal, mechanical, or chemical treatments. The damage induced can bring about their partial or total destruction.

Furthermore, biological materials have specific bonds combining both chemical bonds and spatial conformations. Proteins are complex constructions with specific spatial conformations.

The recognition of a specific site by a molecule enables a highly specific type of bond called an immunological reaction. This type of bond occurs in antigen–antibody reactions. The antigen–antibody reaction (Ag–Ab) is due to the interaction between the epitopes of the antigen and the paratopes of the antibody. The reaction involves four types of non-covalent bonds: hydrogen bonds, electrostatic bonds, hydrophobic bonds, and van der Waals forces.

2.2 Type of Materials and Chemical Bonds

The different types of materials (metal, semiconductor, ceramic, mineral, polymer, biological material, and mixed–composite material) contain one or more different types of chemical bonds or electronic structures, giving them their specific physical properties. These different types of materials can be divided into five classes based on the nature of the chemical bonds of which they are composed.

Metallic Materials: These are pure metals and their mixtures or alloys. They essentially contain metallic bonds. Metallic bonds are characterized by an electronic structure with free electrons in the conduction band and no forbidden bands.

Semiconducting Materials: Their name is directly related to their electronic property, which differs from that of metals by the presence of a forbidden band and an unoccupied conduction band. This electronic structure, or band structure, is at the origin of the effects of impurities or dopants on the electronic, optical, and mechanical properties of semiconductors, which will either raise or lower the Fermi level (the last electronic level occupied by the valence electrons).

Mineral Materials: These are rocks, oxides, mineral glasses, and ceramics. They contain ionic and/or covalent bonds and may be attached to biological materials such as bones, teeth, shells which are of biological origin but with mineral compositions. Ceramics are compounds with covalent and/or ionic bonds that are either natural or synthetic. Among them are the families of oxides, carbides, nitrides, etc., which are either simple or complex. In the oxide family, the structures and properties can vary widely depending on their oxygen stoichiometry. Compounds with mixed cationic valences are very sensitive to the effects of radiation, which could result in the formation of non-stoichiometric compounds.

Organic Matter (Polymers, Biological Materials): This includes C-, H-, and O-based compounds of organic or synthetic origin. These are molecular systems composed of a high number of atoms that contain covalent, ionic, and weak bonds.

Composite Materials: Composites combine different materials at different scales, forming a new material with specific properties. These materials contain a mixture of materials from the four preceding classes.

2.3 Chemical Bonds and Mechanical Properties

The mechanical behavior of materials is related to the nature of the chemical bonds, but especially to crystal structure and organization. This will be an important factor in TEM sample preparation, given the fact that one must go from macroscopic forms to microscopic dimensions by cutting the initial material and/or performing different types of mechanical polishing. The fundamental properties of materials are their rigidity, tensile strength, and plasticity. Material hardness ranges from hard to soft; it is measured by the Mohs scale, which, among other things, characterizes resistance to indentation through pressure. Hardness is a function of the rigidity of the crystal network or the degree of reticulation that exists between the atoms or molecules making up a material. It may be supplemented by the notion of elasticity, measured by Young's modulus. Elasticity is characterized by the material's ability to return to its original shape and volume after a stress is applied. Tensile strength is the evaluation of the material resistance, ranging from brittle to ductile. A material's ductility is its ability to deform under a given stress without breaking. Thus, the breaking point is defined as the resistance to the propagation of cracks before the material breaks. A material is brittle if it breaks easily. Ductility enables the material to be shaped. For softer and therefore more ductile materials, we refer to plasticity.

2.3.1 Mechanical Properties and Crystallinity

A material's fragility and ductility first depends on the interatomic bonds. Not all crystalline materials are ductile: Ductility is a characteristic feature of a deformation produced by the gliding of atomic planes. It involves the presence of crystal defects such as dislocations. These dislocations must not only be able to form but also move easily under the effect of a stress.

In a metallic material, the absence of preferential direction of the bonds between atoms facilitates the displacement of dislocations and will not result in the definitive rupture of the bonds when these dislocations are displaced under the effect of the deformation. The ductility of a metal is even greater if there are a large number of different slip planes.

Thus, the ductility of metals with face-centered cubic (fcc) structures is greater than the ductility of metals with body-centered cubic (bcc) structures, which is, in turn, greater than that of metals with hexagonal close-packed (hcp) structures.

In a material with covalent bonds, which are highly directional, the displacement of the dislocation generally results in a definitive rupture of the bond between the atoms and a rupture of the material along the glide plane. Materials with covalent bonds will be brittle.

In a material with ionic bonds, the bonds are also directional. These materials will display brittle behavior.

In organic materials (most polymers and biological materials), the structure is amorphous, and yet they may be ductile. In these materials, ductility may not result from dislocation displacement, since dislocations may not be present in a non-crystalline material. Elasticity and ductility in these materials are linked to the flexibility of molecular chains and their deformable macromolecular configuration. In soft biological organic matter, the significant presence of water increases the flexibility of the molecular chains and their ability to deform and return to their original shape.

2.3.2 Rigidity: From Hard to Soft

The hardness of a material characterizes its resistance to penetration. It also characterizes the intensity of the atomic bonds, structure, and crystallization. It may be measured by the impression of an indenter of known geometry (tempered steel billet, square-based diamond pyramid, or tempered diamond cone) applied to the surface of a material. The smaller the indentation left behind, the harder the material. On the other hand, a material that is not hard is therefore soft; in other words, it is easily deformable, penetrable, or abraded. While hardness is not a simple property to define, its measurement gives a good idea of the mechanical properties of a material. There are several hardness scales based on the Vickers, Brinell, Rockwell, and Shore measurement tests. The well-known Mohs scale is based on resistance to abrasion: Each body scratches the material below it and is, in turn, scratched by the material above it. This makes it easy to gauge the hardness of a material. The scale is composed of 10 materials ranging from talc to diamond, with 1 denoting minimal hardness and 10 indicating maximum hardness.

For informational purposes, Table 2.1 shows examples for all of the types of materials, which are expressed using the Mohs scale.

Metals have variable hardnesses ranging from very soft (copper) to very hard (certain metal alloys). Semiconductors generally have average hardnesses. Minerals have variable hardnesses. Ceramics are generally harder than the preceding classes. Polymers have widely ranging hardnesses.

Animal biological organic matter, tied to its high proportion of water content, is soft to very soft. Plant organic material, less hydrated and composed of molecules that are often oriented (cellulose and lignin), is harder.

2.3.3 Tensile Strength: Ductility–Brittleness

A material's brittleness derives from the fact that it cannot deform under a stress without breaking due to the rigidity of the crystal lattice. Generally, the most ductile materials are found among metals and alloys. On the other hand, ceramics have ionic bonds and exhibit brittle rupture. For example, materials such as quartz minerals or the ceramics Al_2O_3 and SiO_2 exhibit brittle behavior. Between the two types of materials, semiconductors (whose crystal structure is formed by covalent and therefore directional bonds) have intermediate properties with regard to plastic deformation. For informational purposes, Table 2.2 shows examples of materials that are brittle, of average brittleness, resistant, or ductile.

Despite the presence of strong bonds, polymers are deformable. This is due to chemical compositions that are made up of light atomic elements (C, H, and O), an amorphous structure, and the flexibility of molecular chains. The hardness of polymers can vary widely at room temperature, depending on the glass transition temperature. They have a high elasticity, meaning that they can withstand deformations and may return to their original shape. The same principle applies to non-mineralized organic biological matter; however, this often crystallized matter is brittle.

Table 2.1 Examples of material hardnesses according to the Mohs scale

Type of material	Hardness				
	Very soft Scale 0–2.0	Soft Scale 2.1–4.0	Average hardness Scale 4.1–6.0	Hard Scale 6.1–8.0	Very hard Scale 8.1–10.0
Mineral $\nabla, \oplus$	Talc, gypsum, sulfur, MoS_2, graphite	Calcite, fluorine, ZnS, $CuFS_2$, FeS, PbS, clay	Apatite, feldspar, FeS, TiO_2, mica, opal, Fe_3O_4	Orthose, quartz, topaz, hematite (Fe_2O_3), pyrite (FeS_2), granite, silicate	HgS, corundum, diamond
Metal $\sum$	Mercury, lead, PbSb	Tin, gold, aluminum, silver, copper, magnesium, cadmium, bronze	Iron, nickel, platinum, carbon steel, stainless steel	Titanium, chrome, tantalum, alloy steel	Zirconium, Cemented steel
Semiconductor			Si, Ge	GaAs, InP	
Ceramic $\nabla, \oplus$				MgO, silicide, spinel	Alumina, carbide, boride, nitride
Polymer ∇, Ω	Polymer	Polymer	Polymer	Reinforced polymer	
Biological material $\nabla, \oplus, \Omega$	Cells, wool, cotton	Collagen, fingernail, cartilage, wood	Bone, cellulose, chitin	Tooth chitin	
Mixed–composite material $\nabla, \oplus, \Omega$	Mayonnaise	Graphite polymer	Concrete, Carbon fiber–polymer matrix	Metallic Metallic matrix -ceramic fiber	Ceramic-carbide

$\sum$: metallic bond; ∇: covalent bond; $\oplus$: ionic bond; Ω: van der Waals bonds

Table 2.2 Examples of materials that are brittle, of average brittleness, resistant, or ductile

Type of material	Brittle	Average brittleness	Rupture resistant	Ductile
Mineral $\nabla, \oplus$	Graphite, diamond, glass, quartz, topaz	Calcite, fluorine gypsum, mica, apatite, feldspar		Talc
Metal $\sum$	Gray iron, crude steel	Bronze, carbon steel	Iron, nickel, cobalt, titanium, tantalum, martensitic steel	Lead, gold, aluminum, silver, copper, tin
Semiconductor ∇	GaAs, InP	Silicon		
Ceramic $\nabla, \oplus$	Alumina, MgO		Carbide, boride, silicide	
Polymer ∇, Ω	Thermohardening polymer		Reinforced polymer matrix carbon Fiber composite	Thermoplastic polymer
Biological material $\nabla, \oplus, \Omega$	Bone, tooth, spicule	Fingernail, cartilage, wood		Soft organic matter
Mixed–composite materials $\nabla, \oplus, \Omega$	Concrete		Metal matrix composite, ceramic, carbon and carbon–fiber, metal and ceramic	Polymer–metal

$\sum$: metallic bond; ∇: covalent bond; $\oplus$: ionic bond; Ω: van der Waals bonds

Mixed–composite materials have different mechanical properties than their components. For example, composite materials made from carbon or ceramic fibers and either resin, metallic, or ceramic matrices have increasingly higher mechanical properties, respectively. This property enables these materials to be used under high-temperature and high-pressure stresses (e.g., the conditions required for aerospace materials).

2.3.4 Mechanical Properties of Organic Materials and Glass Transition (T_g)

Chain flexibility is a function of temperature. Therefore, the rigidity and tensile strength of a polymer are also closely linked with temperature. When an amorphous polymer is cooled to below a certain temperature, it becomes hard and brittle like glass. This temperature (which differs for each polymer) is called the glass transition temperature and is abbreviated as T_g. This thermal transition is a material-specific physical characteristic. Hard plastics such as polystyrene and methyl polymethacrylate are used below their glass transition temperature in the vitreous state. Their T_g values are well above room temperature, on the order of 373 K. Elastomeric rubbers such as polyisoprene and polyisobutylene are used above their glass transition temperature, i.e., they are used in the rubbery state and are both soft and flexible. Flexible plastics such as polyethylene and polypropylene are also used above their glass transition temperatures. If the temperature is high compared to the glass transition temperature, the polymer chains can move easily. Below T_g, the chains are not able to move into new positions in order to reduce the stress. A chain that can move easily will result in a polymer with a low glass transition temperature, while a chain that hardly moves will result in a high-T_g polymer.

Glass transition is produced in amorphous polymers. Fusion is produced in semi-crystalline polymers. Fusion is the passage of polymer chains from an ordered crystalline state to a disordered liquid state, but even semi-crystalline polymers contain an amorphous portion, which generally represents between 30 and 60% of the polymer's mass. This is why a semi-crystalline polymer material has both a high glass transition temperature and a high fusion temperature. The amorphous portion only undergoes glass transition and the crystalline portion only undergoes fusion.

For biological materials, generally only the T_g of water is taken into account. At room temperature (298 K), water is always above T_g (138 K) and is therefore in a soft phase. But a biological organism is in fact a heterogeneous compound; certain portions can have an oriented texture (proteins in particular) that makes the whole organism more rigid. Therefore, T_g is not the only parameter that needs to be taken into account. For all composite materials, there will be several T_g values and several components in the overall result of the rigidity or tensile strength.

Conclusion

A material can be both hard and brittle (e.g., some metal alloys, ceramics, or diamond). The hardness of these two cases is different: One is associated with the significant formation of dislocations that form dense lattices (structural hardening of alloys) and makes the material brittle; the other is associated with the nature of the chemical bonds. A material can also be ductile or soft and resistant like certain alloys (shape memory alloys), polymers, and composites. The characteristics of these materials will be important to the use of mechanical thinning techniques on samples. Difficulty will arise when the compound to be studied is composed of different types of materials with substantially different mechanical properties. This is the case with mixed–composite materials.

3 Microstructures in Materials Science

3.1 Problems to Be Solved in Materials Science

Regardless of the formation method used (such as classical sintering or sintering under charge for polycrystalline materials, cathode pulverization, laser ablation for depositing films in thin layers or multiple layers, or the flux method for single crystals), the thermal treatment conditions will be a determinant with regard to the growth mechanisms of the microstructure formed. Thus, parameters such as sintering temperature, controlled oxidation, reduction, neutral atmospheric conditions, heating and cooling rates, mechanical pressure exerted during formation, temperature gradients, and annealing times at high or low temperature all result in the formation of a material that is more or less homogeneous from a textural, structural, and chemical viewpoint, with very specific physical properties.

The formation of most materials involves thermal treatments at relatively high temperatures (573–2,273 K). Material transport via diffusional processes during thermal treatment will lead to the formation of two sorts of microstructural defects during crystal growth at high temperature: point defects (vacancies, interstitials) and extended defects such as twins, dislocations, stacking faults, chemical reactions at the interfaces, secondary solid or liquid phases. Other defects such as microcracks, second-phase precipitations, and phase transformation twins can also be formed during cooling.

Regardless of whether the material studied is a metal, alloy, semiconductor, ceramic, polymer, or mixed–composite material, its formation method will be selected based on the specific physical application desired. In order to respond to this "material problem," it is necessary to characterize the material through observations and different types of analyses. To obtain a given property, a formation method is chosen for study in the TEM in order to determine its properties and structure. To do this, it will be necessary to select the preparation method for its characterization.

3.2 Materials Microstructures

The different types of material microstructures are represented in Figs. 2.2, 2.3, and 2.4. They are classified according to the material's macroscopic organization: bulk (Fig. 2.2), thin layer and multilayer (Fig. 2.3), and fine particles (Fig. 2.4).

Figure 2.2 shows a classification that is arranged by increasing degree of crystallinity, from amorphous materials up to single crystals and bicrystals. What differentiates them from one another is the extent of crystal order. This order is minimal (on the order of interatomic distances) in an amorphous material and increases from poorly organized materials up to textured materials and then to microcrystallized materials (0.1–1 μm), polycrystalline materials (several micrometers), which may be textured, and lastly to single crystals or bicrystals (several millimeters). A microstructural defect in an amorphous material will be crystalline in nature, whereas defects in crystalline materials will be precipitates or amorphous phases, dislocations, vacancies, twins, crystalline or chemical second phases, etc. A single microstructural defect can act effectively for one type of property and be catastrophic for another type.

In Fig. 2.3, we can see that the thin layers (amorphous or crystalline) can be self-supported or on a monocrystalline, polycrystalline, or amorphous substrate. Lastly, Fig. 2.4 shows the different forms of fine particles or single particles.

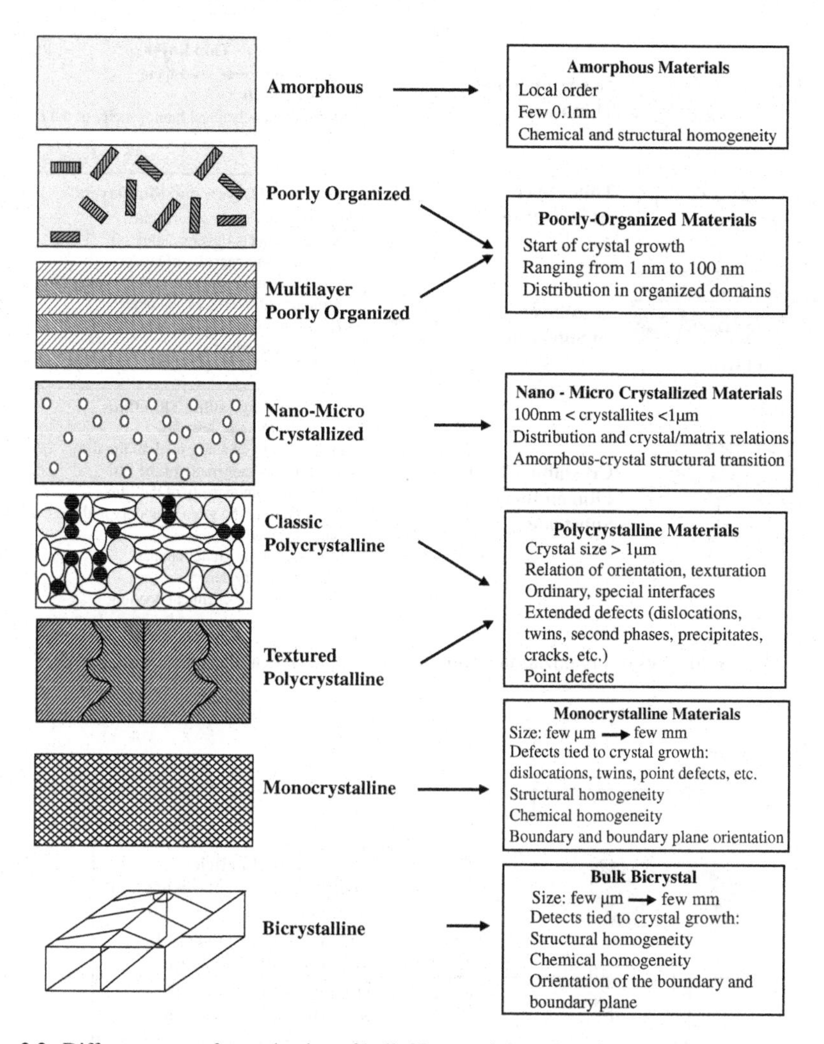

Fig. 2.2 Different types of organization of bulk 3D materials at the microscopic scale

Amorphous materials have a short-range order, which is the interatomic distance. The result is that there is no crystalline order. The analysis of these materials will focus mainly on chemical composition and chemical and structural homogeneity. Crystal growth will be limited to the growth caused by the local crystallization of the nanoparticles present, as can be observed in fluoride glasses, which are used for their optical properties. These are bulk materials and may be either compact or porous.

Poorly organized materials are materials whose crystal growth extends over distances of up to 100 nm. Examples of poorly organized materials often include natural organic materials derived from the alteration of trees, forests, algae, etc.,

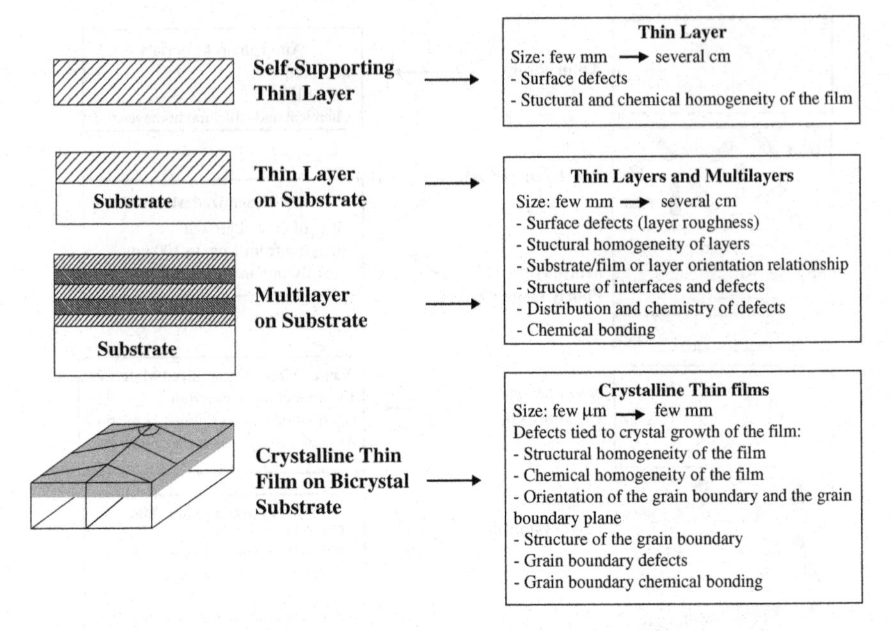

Fig. 2.3 Different types of organization of thin layer 2D materials at the microscopic scale

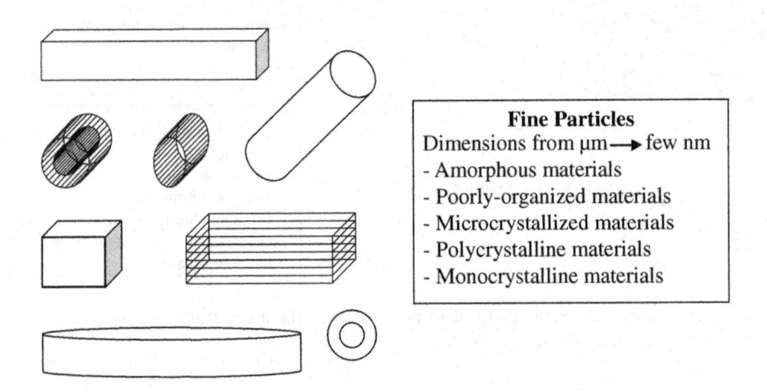

Fig. 2.4 Different types of organization of fine particles or single particles at the microscopic scale

and which are more or less evolved (kerogenes and mesophases), and certain alloys. In this type of microstructure, the interest lies in the start of crystal growth and its location and distribution in the volume of the material, in order to get back to the formation mechanism. Crystalline order may show texturation, such as that present in petroleum source rocks, which is characteristic of certain chemical compositions of C, H, and O in the rocks. These materials can be bulk or fine particles.

Microcrystalline materials are materials in which the order is 3D like that of a crystal. Their dimensions vary from 100 nm to 1 μm at most. In this type of microstructure, there are a large number of interfaces. The crystal orientation varies from one crystal to another and a structural analysis can show either the distribution of single crystals if they are in a matrix or the crystal-amorphous structural transition if the material is manufactured using tempering. This type of structural analysis will provide information on the chemical distribution. These materials are either bulk or fine particles and may be either compact or porous.

Polycrystalline materials are composed of single crystal grains whose size is greater than 1 μm and may extend to several hundreds of microns. In this type of microstructure, it is important to determine whether there are orientation relationships between the grain boundaries, particular interfaces, or other interfaces, as well as to determine if there is texturation. Characterization will also help to determine the formation of (1) secondary phases belonging to the equilibrium phase diagram; (2) particular microstructural defects, such as cavities formed by diffusion mechanisms; or (3) microcracks due to crystal anisotropy and chemistry (possible segregations). These materials are bulk and may be either compact or porous.

Single crystal or monocrystalline materials are crystals whose crystalline order extends over three directions and may range from a few microns to a few centimeters. Their microstructure is linked to their crystal growth, and in particular to their growth mechanism. For example, dislocations, which provide evidence of pyramidal crystal growth on the surface, can be seen inside these materials. Point defects, dislocation loops, and growth or annealing twins may also be detected depending on the structure of the crystal. In certain cases, chemical inhomogeneity can be detected, often in ionic compounds that demonstrate variations in stoichiometry, which can lead to structural variations or to the presence of chemical impurities.

Thin layer or multilayer materials have microstructures with layer surfaces and interfaces oriented on a substrate. Passage from a bulk 3D material to a thin 2D layer material is characteristic of an anisotropy of one of the structure's dimensions in a ratio on the order of 1:10; in this case, it is the direction perpendicular to the substrate. The dimension in the plane of these materials can range from a few millimeters to several centimeters, as with electronic components that are several centimeters in size. These materials can be either amorphous or crystalline. Analyses of their microstructures will include the study of the surfaces, substrate/layer interface, and layer/layer interfaces. In the case of crystalline materials, the epitaxial or heteroepitaxial growth depends on crystalline parameters that define the connections between the atomic planes of the substrate and the layer as well as the connections between the layers themselves. Depending on the lattice mismatch between them, interfacial dislocations will form and lead to coherent, semi-coherent, or incoherent interfaces (depending on the case). The formation of roughness at the interfaces can be quantitatively measured and directly tied to the film's growth mechanism mode (2D, 2D–3D, or 3D) and growth conditions. The analysis of the film in the planar-longitudinal view can be used to analyze the growth islands, while cross sections can be used to analyze the interface structure and chemistry.

3.3 Polymer Microstructures

Polymers (or macromolecules) are molecular systems composed of a very high number of atoms (e.g., 1,000 to 100,000 to 1 million or more). Their molecular masses range between 10,000 and several millions of grams per mole. They are derived from a process called polymerization. In Fig. 2.5, a distinction is made between the following:

(a) Linear polymers, soluble polymers.
(b) Branched polymers, whose solubility is close to that of linear polymers but whose morphology is different.
(c) Crosslinked polymers, which form an insoluble and infusible network (elastomers).
(d) Cyclolinear polymers, which are generally crystalline and have the same properties as linear polymers, except for solubility.
(e) Ladder polymers, composed of linear molecules in which skeletons are attached together in regular sequences. They have good thermal properties but low solubility.

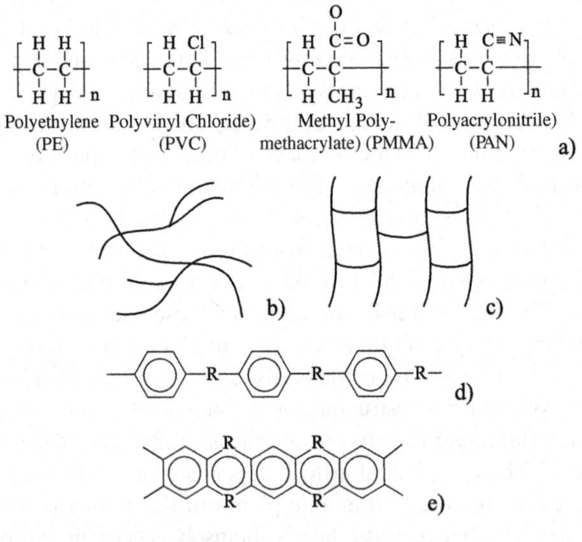

Polyethylene (PE) Polyvinyl Chloride) (PVC) Methyl Poly-methacrylate) (PMMA) Polyacrylonitrile) (PAN) a)

b) c) d) e)

Fig. 2.5 Different structures of polymers

Among polymers, a distinction is made between natural polymers (cellulose, natural rubber from the plant kingdom, silicates from the mineral kingdom, proteins and nucleic acids from the animal and plant kingdoms), artificial polymers resulting from the chemical modification of natural polymers, and synthetic polymers resulting from human engineering.

There are polymers that are potentially soluble, 1D, and thermoplastic, as well as insoluble and infusible 3D polymers with variable crosslink density, often referred to as thermohardening.

Chain Structure and Flexibility: Chain-segment mobility depends on three factors: molecular interaction, steric hindrance, and the number of bonds capable of rotating per unit length of the chain. Intermolecular interactions are secondary bonds of an electrostatic nature: van der Waals forces and hydrogen bonds. The steric hindrance of lateral groups is likely to limit segment rotation.

The study of microstructures will essentially deal with multiphase polymers; it will be concerned with the different domains of these phases and the corresponding interfaces.

3.4 Crystalline Defects and Properties of Materials

Whether the system is bulk, thin layer or multilayer, or single-phase or multiphase, it is the microstructural defects that will give the material its properties. Crystalline defects are atomic arrangement defects of a crystal, formed during growth or through doping or deformation or through martensitic phase transformation. Defects and their interactions play a fundamental role on the macroscopic physical properties of crystalline materials, in proportion to their concentration. Included among microstructural defects are mainly point defects (vacancies and interstitials), linear defects (dislocations), and 2D or surface defects, which are either homophase interfaces (grain boundaries, twin boundaries, antiphase walls, and stacking faults in a single phase) or heterophase interfaces or phase boundaries (defects between phases).

Below is a non-exhaustive summary of the roles the main crystal defects have on the properties of a material.

Point defects are 0D structural defects (vacancies and interstitial). A vacancy defect (Fig. 2.6) corresponds to the loss of an atom in the crystalline network and an interstitial defect (Fig. 2.7) corresponds to an atom occupying an unoccupied site in the crystalline network. These defects are responsible for the properties of transport by means of atomic diffusion in all crystals. The condensation of vacancies or interstitials creates dislocation loops (Fig. 2.8). These defects accelerate the atomic diffusion and kinetics of phase transformations such as precipitation. In ionic crystals and oxides, point defects are electrically charged. They can then trap electrons or holes, resulting in variations in electronic and/or optical properties. Point defects change the levels in the forbidden band of semiconductors, increasing the recombination of carriers and modifying the electronic properties.

Linear Defects: Dislocations are linear structural defects (1D) that result in an additional plane in atomic stacking (Fig. 2.9). They are primarily found in metals and alloys and are involved in plastic deformation. Partial dislocations border an atomic-plane defect and close in on themselves, creating a dislocation loop. The density of dislocations evolves even during deformation; it is able to either increase

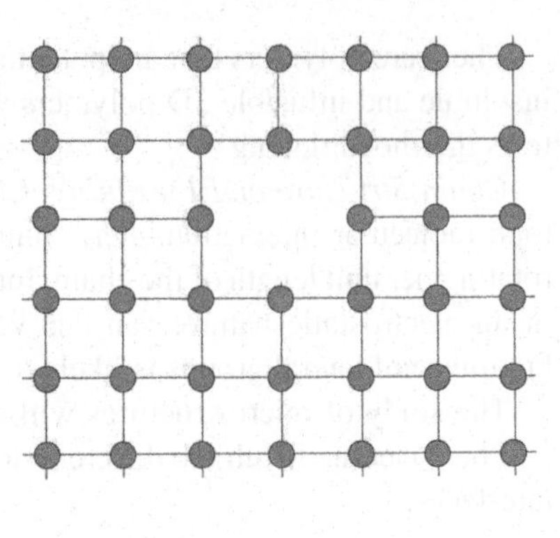

Fig. 2.6 Crystalline point defect: vacancy

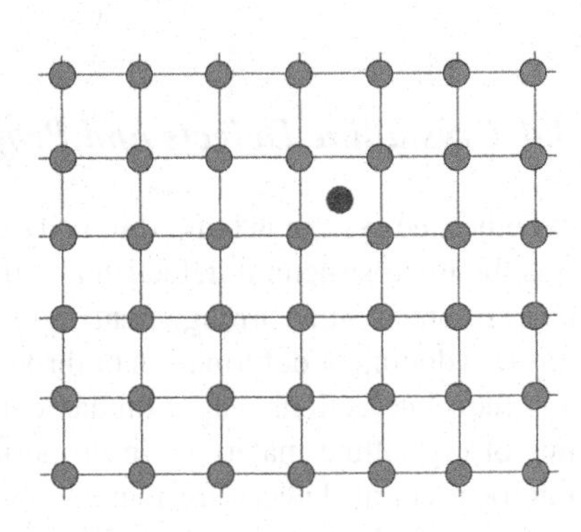

Fig. 2.7 Crystalline point defect: interstitial

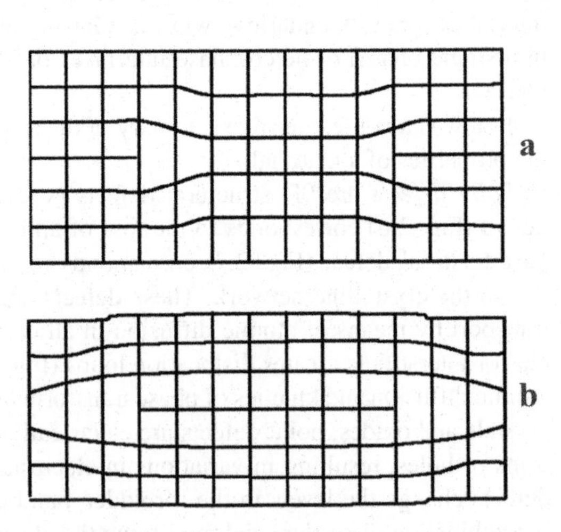

Fig. 2.8 Dislocation loops formed by (**a**) vacancy condensation or (**b**) interstitial condensation

(strain hardening) or decrease (dynamic restoration). Dislocations can either be generated during crystal growth or appear due to mechanical stresses that are either thermal in origin or tied to the presence of composition defects. The occurrence of

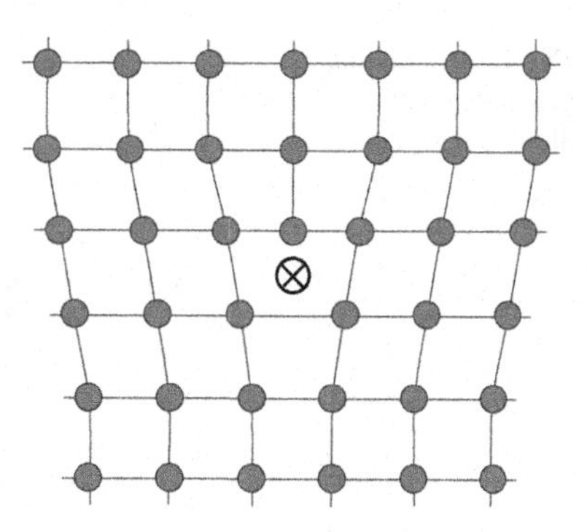
Fig. 2.9 One-dimensional crystal defect: dislocation

dislocations in semiconducting materials is critical, as it severely disturbs the electronic and optical-electronic properties of the components and therefore degrades their performance. Dislocations also introduce levels in the forbidden band.

Grain Boundaries: A grain boundary is a 2D crystal defect.

Its structure will depend on the misorientation between the grains. An example of misorientation is represented in Fig. 2.10. Some examples of grain boundary orientation are represented in Fig. 2.11: symmetric, asymmetric, or faceted. The structure of the boundary plane depends on the growth mechanism, whether the material is bulk and formed in the solid state or formed by deposition in a multilayer material. The misorientation between the grains will define the electrical and electronic conduction properties. Their structure, coherence, possible mobility mechanisms, and possible interactions with other interfacial defects (segregation of impurities, dislocations, and precipitates) have different roles depending on the field of application: catalysis, oxidation, adhesion, crystalline growth, electronic properties, corrosion, etc.

Interfaces: Interfaces can be coherent, semi-coherent, or incoherent depending on the misfit parameters $\Delta a/a$ between the two phases. The interfaces are coherent below the value of $\Delta a/a \leq 1\%$, semi-coherent for values $\leq 5\%$ ($\Delta a/a \geq 1\%$), and incoherent for higher values ($\Delta a/a \geq 5\%$). Figure 2.12 shows two interfaces: one is coherent and the other is semi-coherent. The latter contains interfacial dislocations to compensate for the stresses that exist between the two phases. The structure of the interfaces, coherence, possible mobility and deformation mechanisms, and possible interactions with other defects (segregation of impurities, dislocations, and interfacial distribution) have different effects on their optical, mechanical, magnetic, and electronic properties, depending on their particular field of application.

Antiphase Wall: The structures of these coherent interfaces will not be very active from a mechanical properties viewpoint because they do not correspond to a misorientation of the crystal but rather to a stacking disorder in the single crystal. Nevertheless, the antiphase walls affect the magnetic properties in certain thin oxide layers.

Three-Dimensional Volume Defects are made up of pores or cavities, inclusions, or precipitates whose crystallographic planes are coherent, semi-coherent,

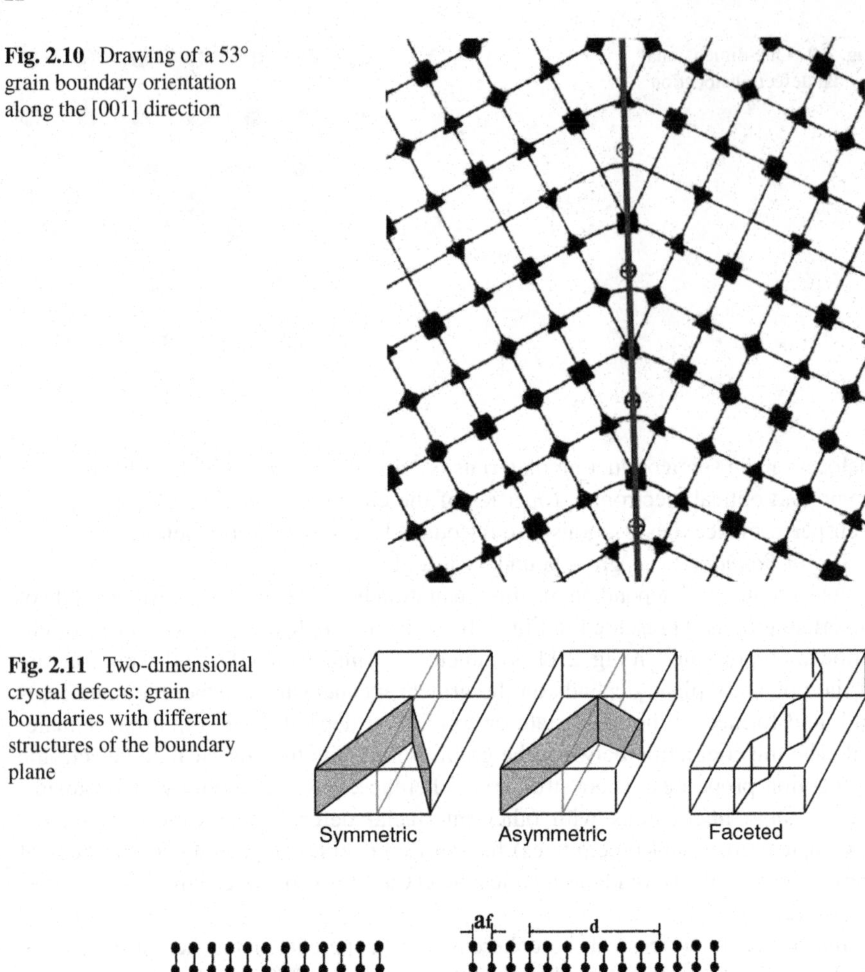

Fig. 2.10 Drawing of a 53° grain boundary orientation along the [001] direction

Fig. 2.11 Two-dimensional crystal defects: grain boundaries with different structures of the boundary plane

Fig. 2.12 Two-dimensional crystal defects: coherent and semi-coherent interfaces

or incoherent with those of the matrix. The accumulation of vacancies can form an intragranular or intergranular cavity.

The mechanical properties of a composite material require good mechanical resistance of the fiber–matrix interfaces under the effects of a load, i.e., during its aging and at high temperatures, which corresponds to the application conditions. Studies on the mechanical behavior of a fiber (e.g., starting with measurements of resistance to rupture) demonstrates that this behavior is directly tied to the behavior of the interfaces found in the composite material. The study of the microstructure consists in determining which defects are responsible for changes in mechanical properties. In the case of a composite, the formation of a second phase at the fiber–matrix interface will enhance mechanical resistance, but may also act as an insulator and be harmful if it is to be used to pass an electrical current. An amorphous phase, even in small quantities such as a thin layer, or microcrystallized precipitates at the level of the interface in a conducting ceramic will be enough to stop the passage of a current. This microstructural defect could be considered beneficial for a particular physical application (e.g., condensers and capacitances) and harmful for another application (e.g., superconducting ceramics) used to pass strong currents. As an example, in the case of thin superconducting monocrystalline layers formed on a monocrystalline substrate, the introduction of grain boundaries makes the value of the critical superconducting current fall by a factor of 100 for films and by a factor of 1,000 for polycrystalline ceramics. Their role is crucial for the electrical properties of a conductive ceramic.

Mechanical, electrical, and electronic properties are tied more to interface structure and chemistry, while magnetic properties are properties intrinsic to the monolithic material.

Crystal Defects in Mixed–Composite Materials: There are a wide variety of mixed materials. In these materials, the mix of material classes gives them multiple properties. The resulting materials are mixed–composite multiphase materials. The presence of more extended, poorly organized, or crystalline domains of different polymers, or of polymers combined with carbon fibers or ceramics as in polymer–matrix composites (carbon fiber, ceramics, etc.), will result in particular high-performance mechanical properties, which are used in the space industry and in aeronautics.

In composite materials, depending on the crystalline nature of the fibers used (carbon, SiC, Al, etc.), the polymer or metallic matrix-based composite material will have mechanical properties that range from high tensile strength to a high Young's modulus. This enables the material to withstand strong mechanical stresses at high temperatures. In the majority of cases, the microstructure of the fiber–matrix interfaces will determine the macroscopic mechanical properties of these materials.

3.5 Solid-State Polymer Properties

Natural polymers (cellulose, proteins, etc.) have a fibrous morphology. The base entity is a fibril just a few nanometers in diameter, with macromolecular chains stretching out along the axis of the fibril.

Most Synthetic Polymers Are Amorphous: Their chains are arranged irregularly. Certain linear polymers are branched but not bridged; they melt or soften upon heating and harden upon cooling. These polymers are referred to as thermoplastic. Other linear polymers undergo structural modifications upon heating that result in irreversible bridging. They then become hard and infusible. This is referred to as thermohardening.

Some Polymers Are Crystalline or Pseudo-crystalline: In crystalline polymers, the lattice parameters are generally between 0.1 and 1–2 nm. In semi-crystalline polymers, the base crystalline entities are the crystalline strip (on the scale of 10 nm) and the spherolite (on the scale of 1 μm). The crystalline strip is a single-crystal polymer in which the length of the chains is folded several times over on itself. A semi-crystalline polymer is a local periodic arrangement of crystalline strips separated by an amorphous phase and connected to one another through binding molecules. The spherolite is a polycrystalline arrangement made up of radial crystallites separated by the amorphous phase, which grows from a center to occupy all available spaces. It has a spherical symmetry. In the case of a constraint (e.g., between two glass slides), growth is 2D and the symmetry is cylindrical. In crystallizable synthetic polymers, the morphologies are essentially lamellar with strip thicknesses on the order of 10 nm.

High rubbery polymers are structured in a 3D network composed of chains connected to one another through functionality nodes. Many polymer materials of this type differ from one another through the chain flexibility and the node type (tangling, covalent bonds, highly cohesive microdomains) and their concentration. This structure gives them particular properties that result in the definition of a new state of matter, the rubbery state, which can be summarized as resulting from certain properties of liquids and elastics. Elasticity is tied to changes in chain conformation. These chain conformations are in fact able to deform between "anchoring points" that constitute the nodes of the macromolecular network.

Today, alloys or blends of polymers are made by mixing thermoplastic components in their melted state in the presence of a third element, which is necessary for developing a morphology. The distribution of components or phases in the alloy will be a determining parameter of the final properties, especially mechanical properties.

The volume fraction of its phases, the morphology or spatial distribution of phases, and interfacial adhesion are characterized.

A particle of any form in a matrix will create stress concentrations that will be favored sites for nonelastic deformations. These deformations will appear as shearing bands (plasticity) or cracks, depending on the polymer considered, and will result in the creation of a core–shell intended to dissipate irreversible energy during an impact.

4 Microstructures in Biological Materials

4.1 Problems to Be Solved in Biology

Living systems are autonomous and self-reproductive chemical "factories" that host thousands of separate chemical reactions that are executed simultaneously and closely coordinated with one another. A large number of mechanisms enable the organism to instantly respond to the internal and external changes to which it is constantly subjected, so as to maintain a remarkably stable equilibrium. As a result of evolution, living organisms (both plants and animals) are very complex and varied. Despite this, the basis of organization on the structural scale remains simple: the cell.

The use of an electron beam propagating inside the vacuum of a TEM represents a dual difficulty with regard to the observation of biological structures: the sensitivity of organic molecules to electron irradiation and the need to eliminate water before observation. These problems require complex preparations, leading to the complete shutdown of the vital process. The challenge with these observations follows the same principle as in any other investigation conducted at any given moment: the researcher obtains precise but incomplete information that does not take constant remodeling into account. Information must often be supplemented by kinetic studies and coupled with biochemical studies. The second difficulty comes from the fact that certain chemical systems do not shut down instantaneously upon the death of an organism. These systems continue to function and they deteriorate the structure before it can be observed in its natural state. In this case, fixation, the step that stops biochemical activities, must be especially fast and effective at the system core in order for it to be studied properly.

Structure has no meaning only insofar as one or more functions associated with it can be recognized. The approach to synthetic phenomena or the degradation of cellular compounds couples observations with labeling-preparation techniques (labeling using gold or platinum particles functionalized for specific enzymatic sites).

In TEM, biological research is conducted at three levels: (1) on the microstructures of protein or nucleic macromolecules; (2) on cell infrastructures, their internal organization, their interconnections, and their relationship with their immediate surroundings; and (3) on the microstructures that correspond to tissue organization.

In the first case, isolation and concentration techniques for a macromolecule made up of cells (proteins or nucleic acids) are used to infer its 3D structure using mathematical reconstruction. Their dimensions are too small to enable them to be characterized directly. This constitutes structural biology.

In the second case, structural recognition is used to deduce their biological functions, metabolic or catabolic activities, and energetic reactions. By investigating these morphologies in the normal state compared to the morphology of pathological tissues, it is possible to deduce the damage caused by these pathologies, define their causes, and select the appropriate treatments. It is also possible to determine, at the cellular scale, the action of medicines used, their sites of action, and possible damaging effects. The study of virus or bacteria microstructures enables an initial approach to determine these microorganisms' families. Recognizing them within a tissue can help to confirm a diagnosis and identify their means of attacking cells, method of replication, and dispersion within the tissue. This constitutes cytology.

A cell must always be considered in its immediate surroundings, i.e., observing the other cells with which it makes contact, thereby creating tissue organizations that have a particular function. It is highly dependent on other cells, which first influence its differentiation, then its activities and therefore its role. All these lead to diverse and varied microstructures. Cells are bathed in a common *milieu intérieur*, with a composition that is regulated for an entire organism. The study of this whole ensemble constitutes a third level of organization that always must be taken into account.

4.2 Singularity of Biological Materials: Importance of the Liquid Phase

Biological material is a multiphase material in liquid solution. Biological structures are composed of carbon compounds such as liquid polymers that are in a liquid solution called the *milieu intérieur*. These compounds are bound to one another by covalent bonds to form methyl, hydroxyl, carboxyl, and amine groups. The cell's small organic molecules have a molecular weight ranging from 100 to 1,000 Da and contain up to 30 carbon atoms. They are usually linear, branched, or crosslinked. They each have a particular spatial conformation that determines their properties. This spatial conformation depends not only on temperature, as with polymers, but also on the water content of their surroundings and its physical characteristics (e.g., pH and ionic composition). Based on their main property, each molecule is grouped together into one of four families of molecules: carbohydrates, fatty acids, amino acids, and nucleotides. Carbohydrates are most often in solution and are not conserved by the preparation methods, with the exception of certain complex polysaccharides such as glycogen, starch, and cellulose. Some amino acids and nucleotides group together to form large polymers called macromolecules, which can be viewed under a microscope. Nonpolar fatty acids such as phospholipids, cholesterol, and glycolipids are amphiphilic. In the aqueous medium of organic tissue, they combine with each other in double layers to constitute membranes or in lipid pools.

Organic matter can be considered as a colloidal solution with an amorphous structure, but the presence of amphiphilic molecules "partition" this medium, causing phase separations and creating compartments. In the living organism, these partitions are in constant motion; the molecules frequently combine and separate very quickly following changes in pH, ionic composition, and water content. Furthermore, the double-lipid layer is crossed by many membrane proteins whose specific functions vary widely. These partitions, which are not at all impermeable, are semi-permeable partitions. They delimit the cells and intracellular compartments that will be described later.

The preparation processes result in the blocking of these molecules, either through freezing while immobilizing the water or through the creation of bridges with a chemical fixing agent, so as to manufacture long insoluble chains that will enable subsequent dehydration. What follows is a denaturing of the proteins and a loss of all of the small diffusible elements. Chemical fixation results in a rigid image of a continuous barrier that has long been attributed to the membrane surrounding the cell, but which is in fact a preparation artifact. In biology, observations are made on an organic material that has been profoundly reworked in comparison to the initial living matter. This gives the material a rigid and identifiable structure, but does not help the microscopist to understand its real structure and remodeling. However, by establishing comparisons between different samples that have been prepared using the same protocol, inestimable information can be deduced, and thanks to labeling techniques, the cell's functions can be determined.

During preparation, the dynamic biogel will become rigid and appear as a multiphase, organized, non-crystalline material. Tissues are considered bulk materials with a liquid solution. With regard to the cells, bacteria, viruses, or macromolecules, they will be considered as suspended sediment. Mineralized tissues may be considered as minerals, insofar as the interest lies only in the matrix and not in the cells that produced it. These tissues are often crystalline or pseudo-crystalline.

4.3 Microstructure in Biology

The very structure of the molecules is problematic given the size of these molecules and their wide dispersion throughout the *milieu intérieur*. In recent years, they have been studied through molecular biology and structural biology. Structural biology reconstitutes the 3D conformation of macromolecules to take account of their biological functions and possibly to modify them. But most often, biologists are interested in a functional approach. Structures are defined according to their role. Looked at from a preparational viewpoint and in comparison with materials, the microstructure of biological materials can be interpreted according to the diagrams in Figs. 2.13, 2.14, and 2.15.

The different types of biological microstructures are classified by degrees of increasing complexity, from materials that can be considered amorphous up to

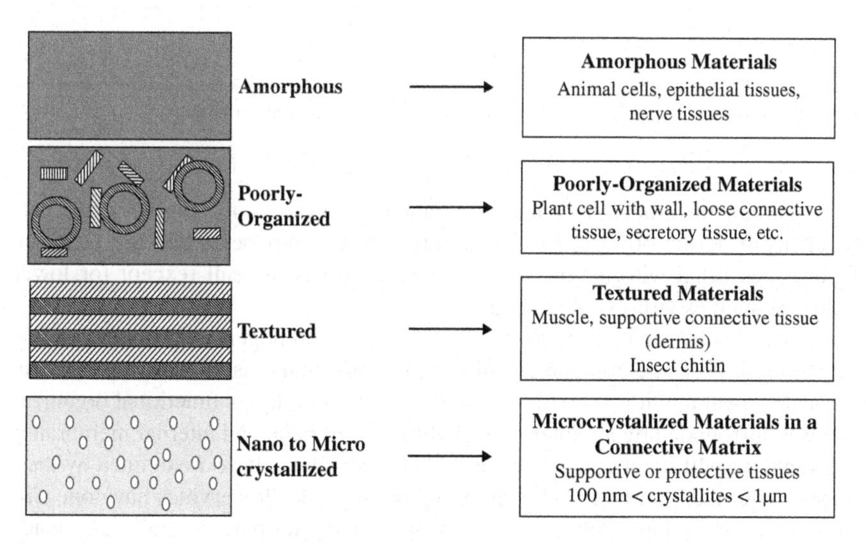

Fig. 2.13 Different types of 3D microstructures of biological materials

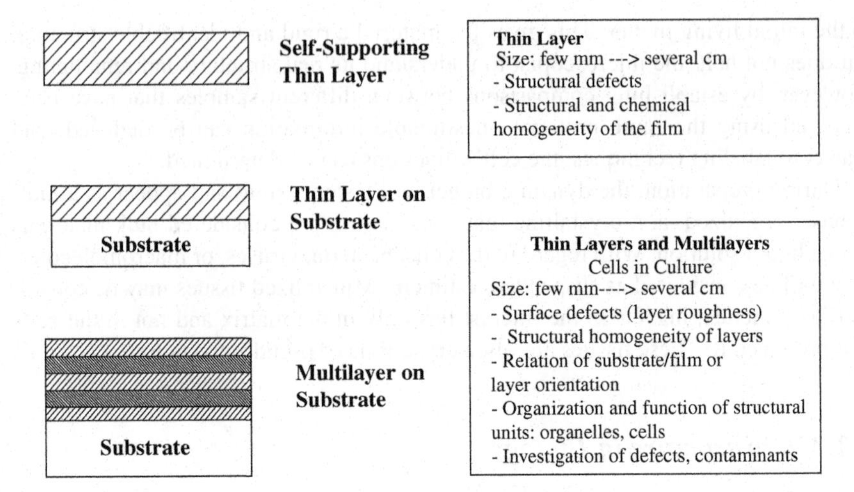

Fig. 2.14 Different types of 2D microstructures of biological materials

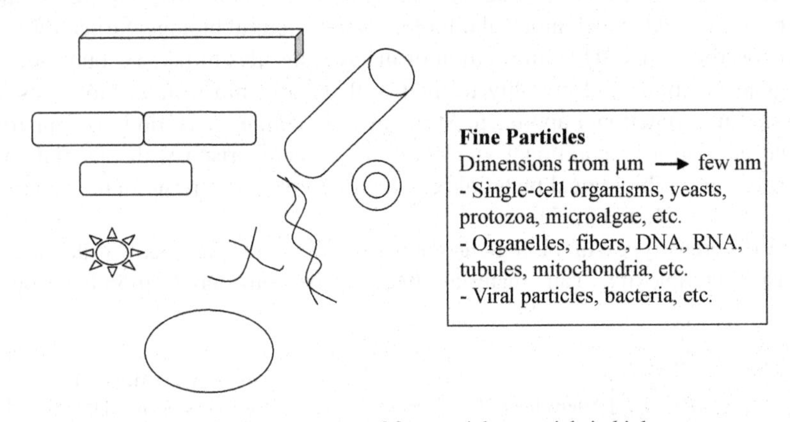

Fig. 2.15 Different types of microstructures of fine-particle materials in biology

mixed–composite structures. What differentiates them from one another is the heterogeneity in terms of hardness or fragility of the components and the presence of porosities filled with air or water. The base unit is the cell (except for lower organisms such as viruses or bacteria).

The animal cell can be considered as an organized homogeneous structure that is non-textured. It can be compared to the amorphous materials of materials science. Physically speaking, it is a hydrated gel. It presents a complex functional organization. It is surrounded by an outer phospholipid membrane, and internal membranes partition it into different compartments called organelles. These are defined by their morphology, size, function, and location within the cell. Prokaryotes have one single compartment (single-cell organisms) while eukaryotes have several. Eukaryotes are often multi-cell organisms. The intracellular space containing the organelles is called the cytosol and it contains many enzymes, ribosomes, and proteins, as well as the cytoskeleton. The main organelles are the nucleus, the Golgi apparatus, the endoplasmic reticulum, and the mitochondria. DNA replication and messenger RNA synthesis occur in the nucleus, the place where the continuity of the cell's different functions is ensured. Ribosomes, which are either bound to the endoplasmic reticulum or not, carry out the replication of proteins. The Golgi apparatus is the site of protein–lipid or protein–sugar coupling and acts as a filter for proteins to direct them toward their final destination. Lysosomes specialize in intracellular digestion. Vesicles ensure the transport of products (either synthesized or degraded) within the cell or to its periphery so that they can be secreted to the cell's exterior. Cellular compartments are connected to each other and two-way exchanges occur.

The plant cell presents a microstructure similar to that of the animal cell, but is distinguished by two essential differences: the presence of a cellulose wall lining the extracellular membrane and the presence of chloroplasts in the aerial parts. The cellulose wall thickens and stiffens the phospholipid membrane. The plant cell presents different hardnesses and fragilities. Furthermore, the wall isolates each cell from its neighbors or from its surroundings. Because of this, each cell can have a different ionic or molar composition than its neighbor, which is not the case with animal cells. The thick walls of plants are true barriers, limiting exchanges with the outside. This poses a problem in that it is difficult for the fixing agent or the inclusion resins to penetrate the cell wall. The chloroplast is the site of photosynthesis.

The plant cell has heterogeneous areas in terms of hardness that are sometimes very impermeable and therefore difficult to embed correctly.

Starting from a simple animal cell, we find increasing complexities of the collections of cells composing tissues.

Epithelial tissues and *nerve tissues* are homogeneous and may be considered as multilayered, often having a polarity (top to bottom).

Muscle tissue is a fibrous and textured tissue.

Connective tissue is a support tissue. It is heterogeneous and more or less textured by the presence of collagen fibers. It ranges from slightly textured to mineralized. It makes up cartilage, bone, dentine, shells, etc.

In plants, the organism's rigidity and support come from lignin, which reinforces the cell wall, particularly in the vessels found in wood. These materials have a poorly organized texture compared to crystalline materials, for example.

Often these different tissues are combined to constitute organs, and the result is a mixed–composite material (bone/cells, epidermis/dermis, parenchyma, and conductive vessel in a plant, for example). Some structures are intended to ensure the continuity of the species and are particularly resistant (e.g., cysts in animals and spores or nodules in plants).

Yeasts are single-cell eukaryotes with a highly individualized nucleus and organelles. *Bacteria* are prokaryotes in which there is no well-defined nucleus or organelles. Yeasts and bacteria have an external membrane that is reinforced by a wall (usually a glycoprotein wall) that enables them to resist harsh external conditions, particularly desiccation. This wall differs in nature depending on the species; it is often difficult to penetrate, making sample preparation for these organisms more difficult. They may be considered to be poorly structured heterogeneous structures.

Viruses are inevitably parasites and are slightly protected by envelopes. Therefore they are very fragile. In particular, they are very sensitive to osmotic or ionic variations that can destroy them. Because of their small size, on the order of some tens of microns maximum, they can all be considered as fine particles in terms of preparation techniques.

4.4 Role of Structures on Functional Properties

The microstructures defined above all perform vital functions for the organism. The development of certain organ functions results in an adapted and often very particular structure. For example, muscle is formed by very large cells that contain myofibrils (actin and myosin) in addition to the organelles mentioned above. These muscle cells are responsible for muscle contraction. Bone favors the construction of an extracellular connective framework that, as it calcifies, gives it rigidity and solidity. These cellular differentiations are defined by the genetic component of the cell and through its relationships with its surroundings. Sometimes these structural modifications are provoked by external aggression (production of collagen in response to a tear or a wound and development of the immune system in response to a viral or bacterial attack). Thus, the absence or exaggerated development of certain organelles in a cell could indicate pathology. An important development in the endoplasmic reticulum results in high synthetic activity, either natural or diverted for the purpose of replicating RNA or DNA for a virus. The disorganization or loss of myofibrils in muscle is the consequence of a myopathy.

Likewise, on the molecular scale, changing the conformation of a protein will reflect different properties. The best known case, because it is the most dramatic, is that of the prion. The presence of certain enzymes on the surface of nucleic acids will enable them to carry out syntheses or replications.

Contrary to materials science, where the structure is considered to determine the different characteristics of a material and therefore its function, in biology it appears preferable to consider the functional role as determining the structure. Through genetic engineering and the creation of the genetically modified organism (GMO), for example, we are starting to see the possibility of intervening on the structure (of DNA or RNA) to change functional properties.

Bibliography

Alberts, B., Bray, D., Lewis, J., Raff, M., Roberts, K., and Watson, J.-D. (1989). *Biologie Moléculaire de la Cellule*, 2e édition. Médecine Sciences Flammarion, Paris.

Ayache, J. (2002). Growth induced microstructure in thin films and heterostructures. In *Crystal Growth in Thin Film Superconductors* (eds. M. Guilloux, V. Perrin, and A. Perrin). Research Signpost, Trivandrum.

Ayache, J. (2006). Grain boundaries in high temperature superconductors. *Phil. Mag.*, **86**, 15, special issue, (eds. P. Bristowe and P. Ruterana).
Bénard, J., Michel, A., Philibert, J., and Talbot, J. (1991). *Metallurgie Générale*, 2e édition. Masson, Paris.
Dorlot, J.-M., Baîlon, J.-P., and Masounave, J. (1986). *Des Matériaux*, 2e édition. Editions de l'École Polytechnique de Montréal, Québec.
Dupeux, M. (2004). *Aide-mémoire Science des Matériaux* (ed. Dunod). Sciences SUP.
Gratias, D., Mosseri, R., Prost, J., and Toner, J. et Duneau, M. (1988). *Du cristal à l'amorphe* (ed. C. Godrèche). Editor « Les Éditions de Physique », Paris.
Guinier, A. and Julien, R. (1987). *La matière à l'état solide*. Liaisons scientifiques Hachette 1987.
Hull, D. and Bacon, D.J. (1984). *Introduction to Dislocations*, 3rd edition. University of Liverpool, UK International Series on Materials Science and Technology.
Putnis, A. and McConnell, J.D.C. (1980). *Principles of Mineral Behaviour*, 1st edition. Geoscience texts, Elsevier, New York.
Rühle, M. and Gleiter, H. (1999). *Interface Controlled Materials*. Euromat 99, Wiley, Weinheim.
Thrower, P.A. (1989). *Chemistry and Physics of Carbons*. Marcel Dekker, New York and Basel.

Chapter 3
The Different Observation Modes in Electron Microscopy (SEM, TEM, STEM)

1 Introduction

Electron microscopy constitutes a key technique for characterizing materials because of its various imaging and spectrometry options. Depending on the scale and nature of the information desired (topographical, morphological, structural, and/or chemical), either scanning and/or transmission electron microscopy is used.

Electron microscopy involves the interaction of accelerated electron source with the sample to be analyzed under vacuum conditions. The resulting interactions cause the emission of various particles or radiation. Collecting these emissions or radiation using different detectors enables the combination of different types of signals with an analysis for the purpose of characterizing the materials. As these detected signals result from different mechanisms of formation, they will provide additional information for characterization.

The microscope vacuum prevents the direct observation of hydrated samples, in particular biological materials, which must be stabilized and dehydrated beforehand.

2 Signals Used for Electron Microscopy

2.1 Electron–Matter Interaction

Electrons from the high-energy incident beam interact with the material and undergo inelastic and elastic scattering, which leads to the emission of electrons and X-rays and light (photons). These radiative emissions can be both low energy (secondary electrons and Auger electrons) and high energy (backscattered electrons). In electron microscopy, the entire range of energies emitted can be explored by

- low-energy transitions of a few electron volts (eV), used to characterize defects and
- transitions of around $\sim$1 keV, used for Auger spectrometry, up to higher energy transitions ($E_0 - \Delta E$), which can be used for the microanalysis of energy losses (Figs. 3.1 and 3.2).

J. Ayache et al., *Sample Preparation Handbook for Transmission Electron Microscopy*,
DOI 10.1007/978-0-387-98182-6_3,

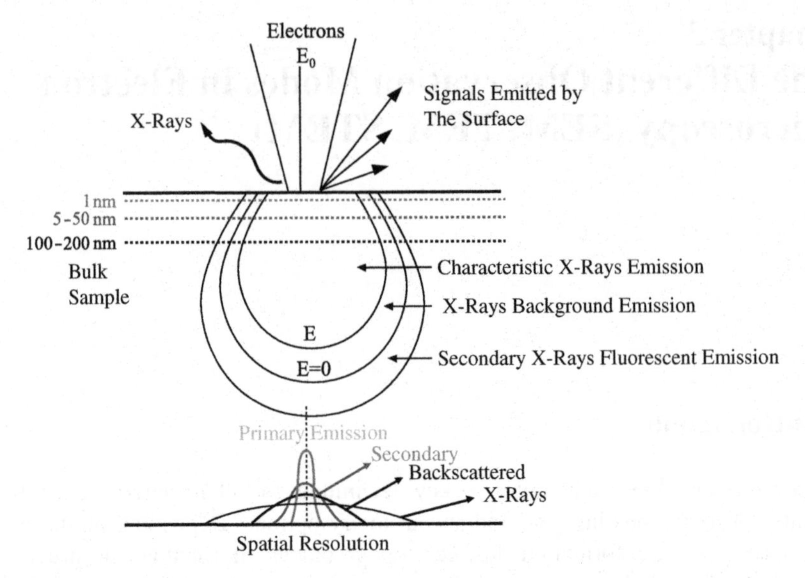

Fig. 3.1 Signals derived from the electron–material interaction within a bulk sample and emitted by the surface

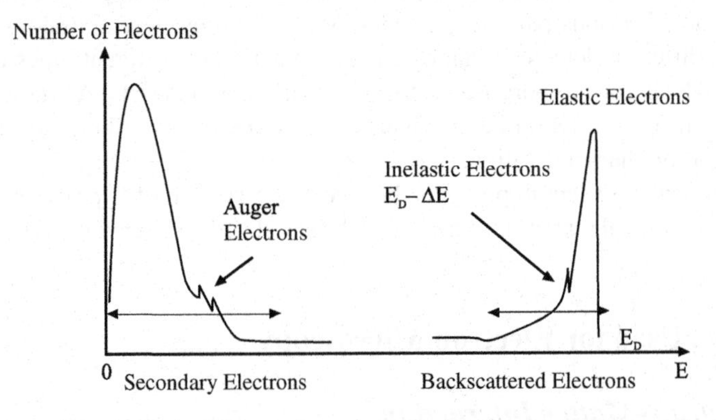

Fig. 3.2 Energy diagram of the different electrons emitted during the electron–matter interaction

Figure 3.1 shows the different signals derived from the electron–matter interaction. The information contained in these signals can be used to characterize materials.

When the sample is thick (Fig. 3.1), the electron beam ends up being absorbed after multiple interactions. Depending on their energy, the primary electrons more or less penetrate the material and gradually lose their energy (through scattering) on the atoms that they encounter and ionize. The result is that other electrons liberated by the sample will in turn ionize other atoms and then scatter in a certain volume called the "excited volume." For example, in nickel, the X-rays emitted come from a volume on the order of 1 μm^3. Secondary, backscattered, and Auger electrons, X-rays, and light emissions generated by the surface will be collected by specific detectors in order to form surface topography images and morphology images and to perform chemical analyses. These chemical analyses can be obtained using a scanning electron microscope (SEM).

2.2 Signals Used for Imaging

Secondary electrons, whose kinetic energy is less than 50 eV, are electrons from the external orbitals of atoms that are ejected by interactions due to incident electrons. Given their low energy, these electrons are easily absorbed by the specimen; only electrons that are able to exit the surface layers (5–50 nm) will be collected by the detector. Their production rate strongly depends on the angle of reflection compared to the primary beam direction, which modulates the contrast. The images obtained will provide information on the sample's surface topography; these images can be produced in an SEM and in a TEM/STEM.

Backscattered electrons, whose kinetic energy is high, are electrons from the incident beam that have undergone few inelastic interactions. They are reflected by the sample. Such a signal contains significant information on the material's chemical composition (*Z* contrast), because the probability of observing such interactions grows considerably with the size of the atom and therefore with its atomic number, and the surface diffraction (EBSD) as well. Because of their energy, the thickness of samples responsible for this emission can reach 200 nm. Depending on how these signals are processed, the resulting images will provide either an image of the elements' distribution according to their atomic number, the surface topographical information, or crystallographic orientation of the surface.

Cathodoluminescence is the emission of photons in the energy spectra corresponding to ultraviolet, visible, and infrared light from an insulating or semiconductor material subjected to electron bombardment. Cathodoluminescence corresponds to the recombination of electron–hole pairs created by the electron beam. These images provide information on the recombination centers, which can be intrinsic or extrinsic (caused by defects). This type of image is also obtained using an SEM.

Electron beam-induced current (EBIC) corresponds to the same interaction mechanism as cathodoluminescence, but also requires the presence of an internal electrical field. The images obtained provide information on the distribution of the electrical properties of the defects that are the recombination centers. This technique is therefore only applicable to semiconductor materials and can be used in an SEM or a TEM/STEM. If there is no internal electric field, an "absorbed current" image is produced. The distribution image of the conductive zones follows the following distribution equation:

$$\mathbf{I_i}(\text{incident el.}) = \mathbf{I_a}(\text{absorbed el.}) + \mathbf{I_b}(\text{backscattered el.}) + \mathbf{I_s}(\text{secondary el.})$$

2.3 Signals Used for Chemical Analysis

X-ray photons are characteristic of the chemical elements present in the sample and correspond to the relaxation signal of the electrons in the internal layers. They are induced by the interaction with the incident electrons. The output of this emission is proportional to the atomic number. The emission, highly dependent on the absorption of the signal by the matrix, is accompanied by a secondary fluorescent emission that will disrupt the quantitative chemical analysis. Software must be used in order to correct these effects for quantifying elements (ZAF correction, see Section 5.5). Imaging is used to conduct quantitative or semi-quantitative mapping of element distribution. Recent developments make it possible to conduct analyses of distributions in the material; they are measured using energy-dispersive spectrometry (EDS) or wavelength-dispersive spectrometry (WDS). Compared to energy-loss analysis (EELS), X-ray spectrometry is more sensitive for heavy elements than for lighter elements. X-ray spectrometry can be performed in an SEM, a TEM, a TEM/STEM, and a STEM.

Inelastic transmitted electrons (EELS or PEELS) are scattered by the object with high energies (near the incident energy $E_0 - \Delta E$). This electron energy loss is characteristic of the element being analyzed. The yield of this emission is more effective for elements with a low atomic number and is less effective for heavy atoms. Energy windows (which limit the range of the energy losses analyzed) may be used to conduct chemical mapping. Therefore, this type of analysis complements chemical analyses using X photons and can be performed in the TEM, TEM/STEM, and especially in a STEM.

Auger electrons have energies just above that of secondary electrons (less than 2 keV). They come from deep, low-energy electron layers in surface atoms. They are emitted through the same mechanism as X-ray photons, minus the ionization energy. This emission will cover a thickness of two to three single layers of atoms on the surface of a sample. Emission probability depends on the atomic number; it is more effective for low-Z values. These electrons are used to make spectroscopic surface analyses in an Auger spectrometer. This process is similar to that of an SEM, but the Auger spectrometer uses a high vacuum (10^{-8} Pa) to prevent any surface contamination since the top atomic layers are being analyzed.

Inelastic transmitted electrons (EXELFS and ELNES): An atom's electrons will undergo chemical displacements of the absorption threshold depending on their chemical environment. Investigation of the threshold structure is used to determine the nature of the chemical bond. Investigation of the modulations present after the absorption threshold is used to determine the number of close neighbors as well as their distance to the atom studied (EXELFS and ELNES). This technique can be used in a STEM as well as in a TEM/STEM. The analysis of plasmons, which corresponds to the analysis of the vibration frequencies of either a crystalline or molecular compound, helps to determine the local dielectric properties of the phase and its concentration.

2.4 Signals Used for Structure

2.4.1 Transmitted Electrons: Thin Samples with Thickness <100 nm

When the sample is thin, the energetic electrons will cross over the sample after undergoing elastic scattering without energy loss and inelastic scattering with energy loss. A small portion of them are absorbed. Elastically transmitted electrons are used by the TEM (Fig. 3.3) to produce structural imaging, crystallography, and spectroscopy (see Section 5). Inelastically transmitted electrons are used to perform various chemical analyses.

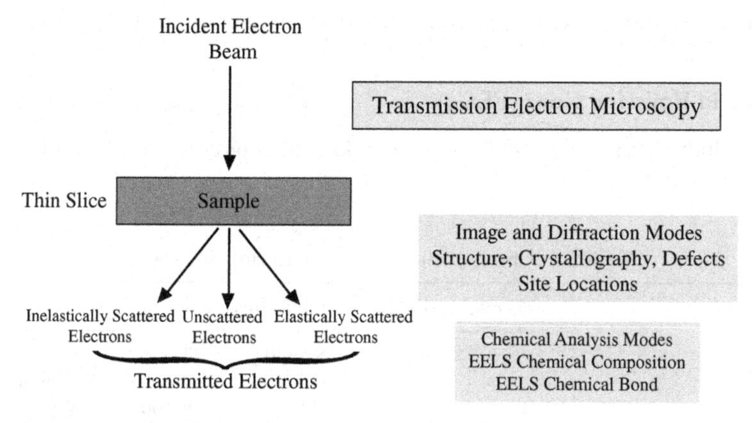

Fig. 3.3 Signals resulting from the electron–matter interaction with a thin sample in transmission electron microscope

After elastic scattering (E_0 energy), the incident beam may either be diffracted according to Bragg's law ($2d \sin \theta = n\lambda$) by the atomic planes of the crystalline material or scattered by an amorphous material.

The interference between scattered and diffracted electrons coming from the object interacting with the incident beam through an electromagnetic lens results in the formation of images containing structural information on the specimen down to the atomic or molecular scale. These images can be obtained using a TEM, a TEM/STEM, and a STEM.

3 Microscopes and Observation Modes

3.1 Illumination Sources

An electron gun is composed of a hot cathode at zero voltage, placed in a tube with a voltage varying around –50 eV in order to regulate the beam intensity (Wehnelt cylinder).

3.1.1 Thermionic Sources

These are made from V-shaped metal filaments or tips composed of a single crystal heated to a high temperature. The electrons extracted from the tip are accelerated toward the sample by the electrical field between the cathode and the anode. Generally speaking, the thermionic gun does not provide a very coherent beam. This is due to the fact that the speed, and therefore the kinetic energy, of the emitted electrons follows a Gaussian distribution, resulting in chromatic aberration.

Tungsten filaments ($T = 3{,}346$ K) and lanthanum hexaboride (LaB_6) with ($T = 1{,}346$ K) tips require vacuums between 10^{-3} and 10^{-7} Pa. Lanthanum hexaboride provides better coherence, resulting from the decreased source size. An LaB_6 filament is commonly used for conventional electron microscopy (Table 3.1).

3.1.2 Field Emission Guns (FEGs)

FEGs include thermally assisted, cold-cathode field emission guns, Schottky effect and hot cathode guns.

Table 3.1 Characteristics of different electron sources

		Thermal emission		Field emission		
		W	LaB_6	Schottky type ZrO/W	Thermal emission W(100)	Cold emission W(310)
Brightness: 5 × 10^5 ($A/cm^2/sr$) at 200 kV		1	10	1,000	1,000	1,000
Source size		50 μm	10 μm	0.1–1 μm	10–100 nm	10–100 nm
Energy window (eV)		2.3	1.5	0.6–0.8	0.6–0.8	0.5–0.7
Operating conditions	Vacuum (Pa)	10^{-3}	10^{-5}	10^{-7}	10^{-7}	10^{-8}
	Temperature (K)	2,800	1,800	1,800	1,600	300
Emission	Current (nA)	1–10	50–100	1–10,000	1–10,000	1–10,000
	Short-term stability (%)	1	1	1	5	7
	Long-term stability (%/h)	1	3	1	6	5
	Current efficiency (%)	100	100	10	10	1
Life span (h)		120	1,000	>3,000	>3,000	>3,000

The cold-cathode gun consists of an extremely acute monocrystalline tungsten tip. It is not heated (pure field effect), but rather a significant electrical field is applied (2–7 keV). These guns require an extremely high vacuum in the microscope (10^{-7}–10^{-8} Pa) in order to prevent tip contamination. This particular requirement makes FEGs very expensive and delicate machines. The emitted electrons have a very low energetic dispersion (see energy window in Table 3.1). The beam is highly coherent.

Another type of gun is composed of a semi-conducting tip (Schottky effect). These guns have very high brightness. These types of guns are used in SEMs and TEMs to produce very bright nanometric-sized probes.

3.2 Illumination Modes and Detection Limits

By controlling the stability of the high voltage and through improvements in objective lens quality, detection limits are related mainly to beam coherence and brightness. In particular, a gun's brightness increases from a source with a tungsten filament to a field emission gun ranging from 10^5 to 10^8 A/cm^2/sr. This increase in brightness enables a reduced spot size, i.e., it reduces the spatial resolution of the analyses. For example, the smallest analyzable size is from 50 to 0.025 μm in an SEM, from 2 to 1 nm in a TEM (from tungsten guns to FEGs) and from 1 to 0,1 nm in a STEM. Additionally, beam coherence is well adapted for high-resolution microscopy conditions (HRTEM).

Lens manufacturing defects must be considered in addition to the type of detection limit described above. These lens defects create an imperfect lens, and as a result they do not have rotational symmetry. Thus, they introduce an aberration for which the microscopist must compensate. The most significant aberration, "the spherical aberration" (C_s), affects the objective lens. This aberration has a defined value for a given objective lens. The spherical aberration astigmatism correction is the determining limitation for the HRTEM imaging illumination mode. Its effect may be reduced by the introduction of an objective aperture or it can be eliminated using C_s correctors. These types of correctors, composed of additional magnetic lenses, are part of the latest generation of microscopes.

3.3 Microscope Resolutions and Analysis

3.3.1 Resolution Limit of the TEM

A microscope's resolution corresponds to the shortest distance between two details that can be distinguished within the image. For a transmission electron microscope, two resolutions are defined: one in two directions, called "point" resolution, which is necessary for atomic resolution, and the other in one single direction, or "line" resolution of the interreticular planes, which is found in the lattice-fringe technique.

The practical electron microscope's resolution in the weak phase object CTEM conditions, is determined using the formula

$$R_p = 0.65(C_s\lambda^3)^{1/4}$$

where C_s is the spherical aberration coefficient (characteristic of the objective lens) and λ is the electron wavelength (a function of the accelerating voltage). Resolution increases when C_s decreases. For example, for a 200 keV transmission electron microscope, a C_s value of 2.2 mm corresponds to a point-resolution limit of 0.27 nm and a line resolution of 0.14 nm, whereas a value of 1.2 mm will change them to 0.24 and 0.10 nm, respectively. Resolution increases with high voltage and changes for the same C_s value (1 mm) from 0.22 to 0.16 nm when going from a 200 keV microscope to a 400 keV microscope.

The objective lenses of analytical microscopes have a higher C_s than high-resolution microscopes because of the space needed to tilt the sample (tilt angles from ±30° to ±70° for the former and from ±10° to ±25° for the latter). The objective aperture makes it possible to eliminate the effect of spherical aberration, but also limits the resolution.

3.3.2 Spatial Resolution

Spatial resolution corresponds to the dimensions of the detected signal's emission zone. It is different for X-rays, secondary electrons, backscattered electrons, and transmitted electrons. This resolution involves the instrument, the sample, and the nature of the signal collected. The lower limit of this resolution depends on the probe diameter, i.e., the diameter of the incident beam at the crossover (or spot mode), which is a function of

- probe current: the higher the current, the larger the probe;
- electronic optical aberrations;
- accelerated voltage, E_0: at equal probe currents, the probe diameter increases if E_0 increases; and
- scattering within the sample (broadening of the beam).

In the case of thin samples (maximum thickness of 100 nm), in an initial approximation spatial resolution corresponds to the size of the probe used for characterization. This is especially true if the sample is very thin (less than 50 nm).

Spatial resolution is also tied to the dimensions of the area the signal comes from; this depends on the signals used. Thus, for secondary electrons the emission area is superficial and spatial resolution is limited by the size of the incident probe (3–5 nm). Spatial resolution decreases for backscattered electrons, which come from a deeper zone (100–500 nm), and then lastly, it is the smallest for X-rays (1 μm).

4 The Different Types of Microscopes: SEM, TEM, and STEM

Modern microscopes are increasingly enabling conventional transmission electron microscopy (CTEM), scanning electron microscopy (SEM), and the scanning electron mode (TEM/STEM) in transmission electron microscopes (TEM).

4.1 Scanning Electron Microscope (SEM)

In the scanning electron microscope (in reflection), the narrow incident electron beam scans the specimen surface in convergent mode by means of the sequential displacement of the electronic probe on the sample (by using the deflection coils). This displacement is combined with the simultaneous synchronized transmission of the signal coming from the sample onto a display screen (as in a television). Each point on the screen corresponds to a point of the probe on the sample. The signal is collected by detectors located above the sample. This signal can be quite varied (e.g., secondary, backscattered, and scattered electrons by the surface, X-rays emitted, the sample current) Therefore, for a chosen signal, the image is a map of the sample's response to the electron probe. The advantage of certain newer SEMs is the possibility of working at controlled pressure near atmospheric pressure or in either humid or reactive atmospheres. SEMs with field emission guns have enabled the use of very small probes and especially low voltages (less than 500 eV) for the direct observation of insulating samples.

4.2 Conventional Transmission Electron Microscope (CTEM)

The column of a transmission electron microscope is composed of a set of electromagnetic condenser lenses that illuminate the sample, according to different modes (parallel or convergent beam), an objective lens to form the image and diffraction of the object, and magnification lenses (diffraction, intermediate, and projector lenses) placed after the objective lens to magnify either the sample's image or the diffraction pattern. The sample is placed in the air gap of the objective lens in a sample holder that can be tilted in one or two directions and possibly rotated. The space available for the sample in the specimen holder is small, on the order of the magnitude of the objective lens's focal length. This lens will determine the type of stage, which can be either high resolution or analytical, enabling tilt angles of $\pm 10°$ to $\pm 25°$ for the former and $\pm 30°$ to $\pm 70°$ for the latter. The microscope's vacuum is about 10^{-5}–10^{-6} Pa, depending on its various pumping systems.

Figure 3.4a shows a cross section of a conventional transmission electron microscope with a symmetrical objective lens that is used to produce a parallel or convergent beam with very small probe sizes.

The second type of microscope (Fig. 3.4b) contains a Köhler illumination system that enables a completely parallel observation mode for all magnifications and an Omega electron energy filter in the column.

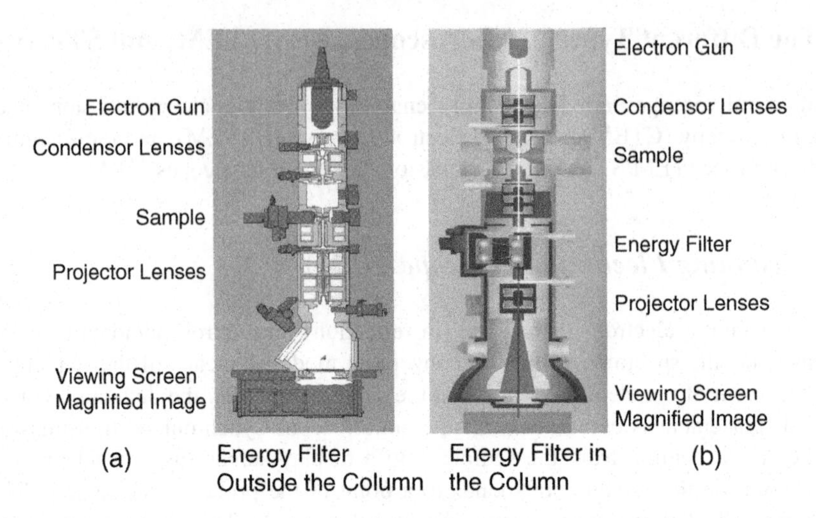

Fig. 3.4 (**a**) Conventional (CTEM) microscope and (**b**) Zeiss TEM microscope with an Omega energy filter in the column

In a conventional microscope, the image is formed by the recombination of diffracted beams coming from the object through the objective lens. Image formation occurs in the Gaussian plane, and the diffraction pattern is located at the Abbe plane (focal plane or diffraction plane) of the lens as shown in Fig. 3.5.

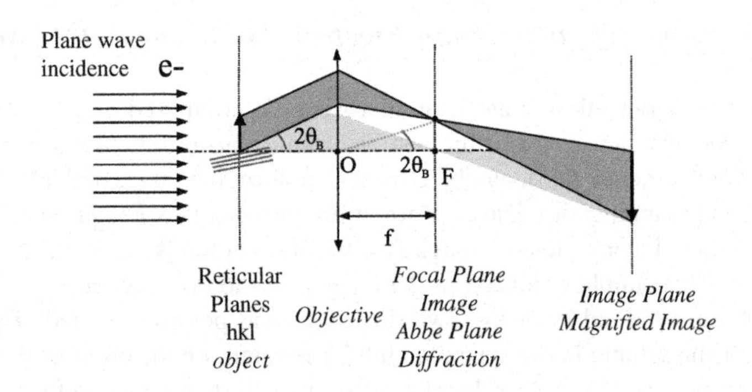

Fig. 3.5 The electron beam's path through a convergent lens

The transmission electron microscope's particularity, compared to the optical microscope, is that the set of magnification lenses can serve to magnify the image plane or the diffraction plane. Therefore, it is possible to change from image mode to diffraction mode by means of a simple commutation that corresponds to different currents of the projector lenses. Thus, there are three image modes: low-magnification mode, high-magnification mode, and selected area diffraction mode, corresponding to intermediate magnification. The diffraction modes include

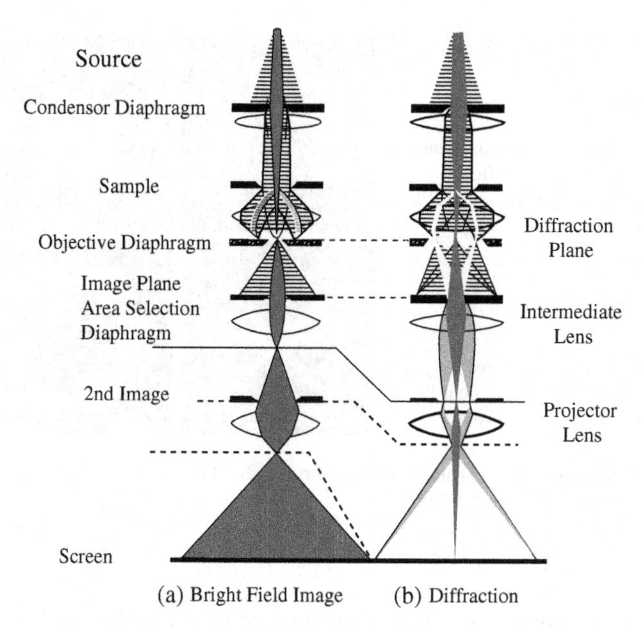

Fig. 3.6 Diagram of the path of electrons through the system of lenses in the TEM column illustrating (**a**) image and (**b**) diffraction modes

selected area electron diffraction (SAED) and convergent beam electron diffraction (CBED). Figure 3.6 shows a comparison of BF imaging mode with SAD mode.

In image observation (MAG mode), the magnified images are formed at different levels in the column, while in selected area mode, both the image provided by the objective lens (which is in the Gaussian plane) and the image of the selected area aperture coincide in the same plane. In selected area electron diffraction mode, the area selected by the aperture in the image plane will provide the diffraction pattern at the back focal plane of the objective lens.

In CTEM mode, the parallel electron beam is broad and illuminates the entire sample at the same time. This allows for a number of imaging modes. Sample illumination in parallel mode is used to make bright-field contrast images (bright-field mode), crystallographic dark-field images (dark-field and weak-beam modes), selected-area electron diffraction (SAED), micro- and nano-diffraction, and high-resolution images (HRTEM). It is also used to perform point chemical analysis using X-rays (EDS) or inelastic electron energy loss (EELS). This illumination mode provides the best observation conditions for high-resolution (HRTEM) or low-dose radiation illumination, as well as for dark-field and weak-beam images.

In TEM mode, a convergent beam (Fig. 3.7) is used to perform convergent beam electron diffraction (CBED), microdiffraction, and large-angle convergent beam electron diffraction (LACBED). It is also used for annular dark-field mode, either

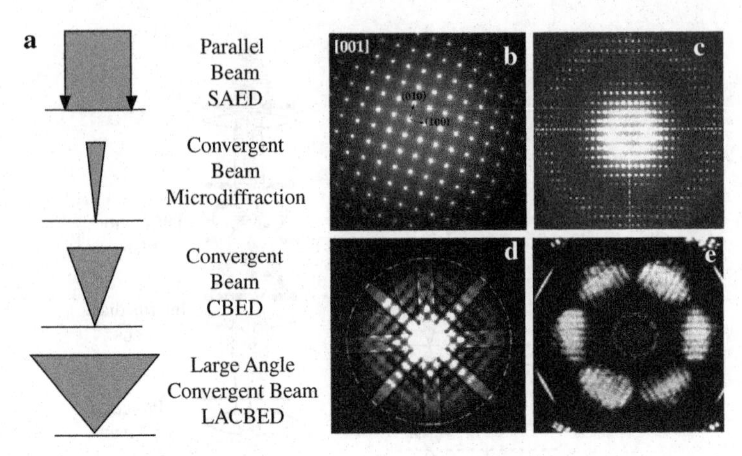

Fig. 3.7 Incident beam modes (a) and type of electron diffraction patterns: (b) SAED, (c) microdiffraction, (d) CBED and (e) LACBED

using a set of incident beam deflection coils or using an annular aperture located at the illumination condenser lens C2.

4.3 Analytical TEM/STEM Microscope and "Dedicated STEM"

The most high-performance equipment for microanalysis is the "dedicated STEM," the only example of which is the HB5 VG (vacuum generator). The vacuum is very high (10^{-6} Pa in the column and 10^{-9} Pa in the gun) in this device in order to help prevent contamination of the tip and the sample. Its particularity is that it has a FEG that emits a narrow beam that scans the sample as in an SEM. STEM imaging only uses transmitted signals. With a narrower size and greater intensity, the emission tips of these types of guns are the only ones that enable analyses at the atomic column scale.

In STEM mode (a narrow, convergent beam scanning the sample), various signals emitted by the sample can be used (e.g., elastic, inelastic, scattered, and non-scattered transmitted electrons, as in Fig. 3.8a) and can be detected using an annular dark-field detector located below the sample.

Depending on both the position of the detector and the detection angle, bright-field or annular dark-field images (ADF and HAADF) can be formed. STEM mode allows for nanodiffraction, EDS microanalysis down to the nanometer scale, and high-resolution Z-contrast chemical imaging, which is also called high-angle annular dark-field imaging, HAADF (Fig. 3.8b). This mode also allows for the analysis of the concentration profiles in EELS image-spectrum mode. One of the advantages of using TEM/STEM instead of a dedicated STEM is the rapid change of the sample without heating or cleaning it by plasma cleaner. The advantage of STEM mode is

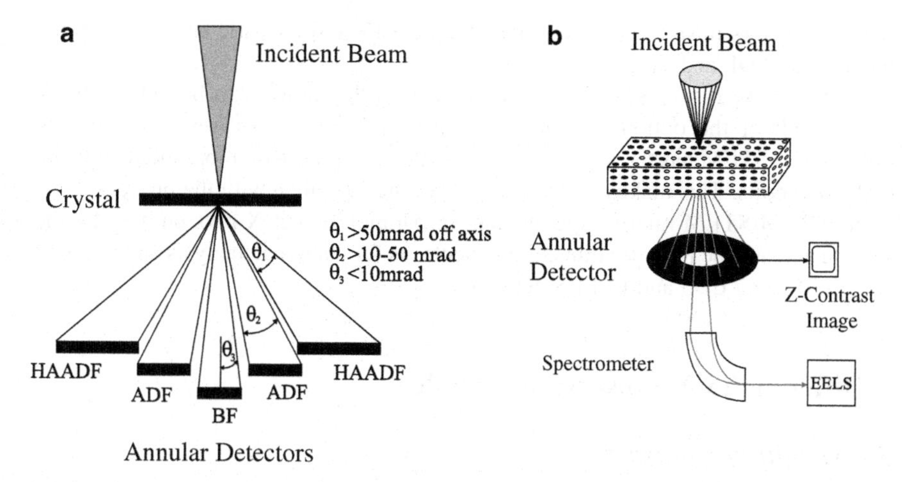

Fig. 3.8 (**a**) Different detectors in STEM mode and (**b**) STEM mode of elastic and inelastic transmitted electrons: ADF or HAADF detector

obtaining a very narrow beam on a thin specimen and the high-energy resolution for EELS analyses. The spatial resolution reached is of the same size as the probe used.

In a "dedicated STEM," as well as in the latest generations of analytical microscopes (TEM/STEM) equipped with field emission guns (FEGs), the electron beam makes high resolution possible because of the very small probe size it can attain (~0.14 nm). It is then possible to analyze and characterize very small areas a

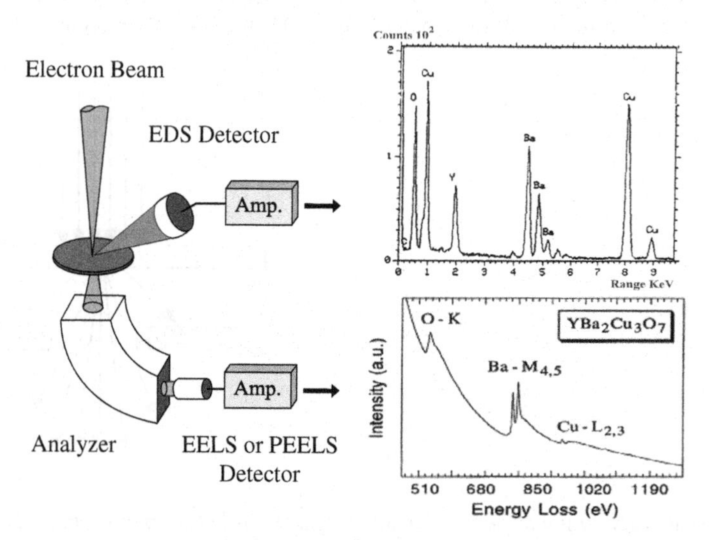

Fig. 3.9 Different types of detection in a TEM/STEM microscope: analysis of X-rays and elastic and inelastic electrons transmitted constituted by a PEELS energy-loss filter outside the column

few nanometers in size using the transmitted elastic and inelastic electron imaging modes (HAADF and EELS).

The purpose of the STEM as an accessory to the TEM is to be able to bring together all of the options for analyzing a sample with good resolution in one device. The STEM has the same functions as the TEM but with more incident beam deflection coils. This makes it possible to scan the specimen with the probe.

In a TEM/STEM microscope we can, in parallel, detect X-rays emitted by the sample and use the inelastic transmitted electrons for chemical analysis and imaging (Fig. 3.9) using EDS and PEELS detectors.

5 Different TEM Observation Modes

5.1 *Origin of Contrast*

The electron beam is characterized by a wave–particle duality of the electrons. The passage of the beam through the sample results in electron transmission deficiency and a phase change of the exit wave. Depending on the type of sample, one of two phenomena is predominant. In amorphous samples, the absorption phenomenon is dominant, whereas in crystalline samples, phase change is essential.

Absorption Contrast (Fig. 3.10): Elastic diffusion electron transmission increases the opening angle of the incident electron beam, which is initially parallel. By placing a so-called contrast aperture at the exit surface of the specimen the extremely high-angle scattered electrons are eliminated, keeping only the electrons that are transmitted practically without scattering. It is then possible to discriminate two points on the sample that have different scattering powers (either the nature of the constituent atoms or the number of these atoms and consequently the local thickness).

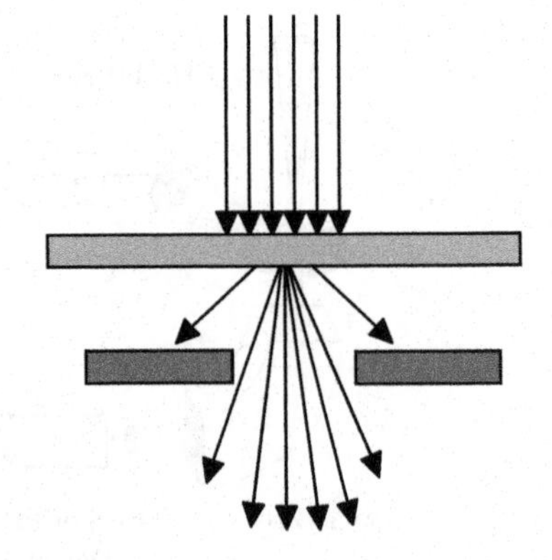

Fig. 3.10 Introduction of contrast using an objective aperture

The basis for viewing structures in any "amorphous" specimen is absorption contrast. An example of this includes samples of polymers or biological materials that are essentially composed of light carbon, hydrogen, oxygen, and nitrogen atoms. In this case, the cellular structures are made visible through selective labeling using heavy atoms (osmium, tungsten, and uranium).

5.1.1 Amplitude Contrast and Phase Contrast

In the case of diffraction, the primary beam must be described as a wave having an amplitude, a phase, and a wavelength. The incident electron beam undergoes phase change by interacting with the atoms, the scattering centers in an amorphous material, or by the crystalline potential of organized material. The incident electron beam may be considered as a plane wave. It is the phase change of this wave by the object that creates the image contrast. At the sample's exit surface, this complex exit-wave function carries the specimen's phase and amplitude information. The contrast observed in microscopy corresponds to the intensity mapping as a function of the x- and y-positions in the image. It is the result of electron scattering by the atoms in an amorphous material and the diffraction of incident electrons by the atomic lattice planes in crystalline materials.

Amplitude contrast corresponds to a scattering or diffraction contrast that depends on each atom's scattering coefficient. For amorphous materials, amplitude contrast is mainly due to absorption contrast and thickness variations (Fig. 3.11a). For crystalline materials, it is due to diffraction contrast by the atomic lattice planes that diffract according to Bragg's law (Fig. 3.11b). In both cases, the contrast is made using an objective aperture that selects electrons scattered or diffracted, in the back focal plane of the objective lens. Without an objective aperture, all the beams

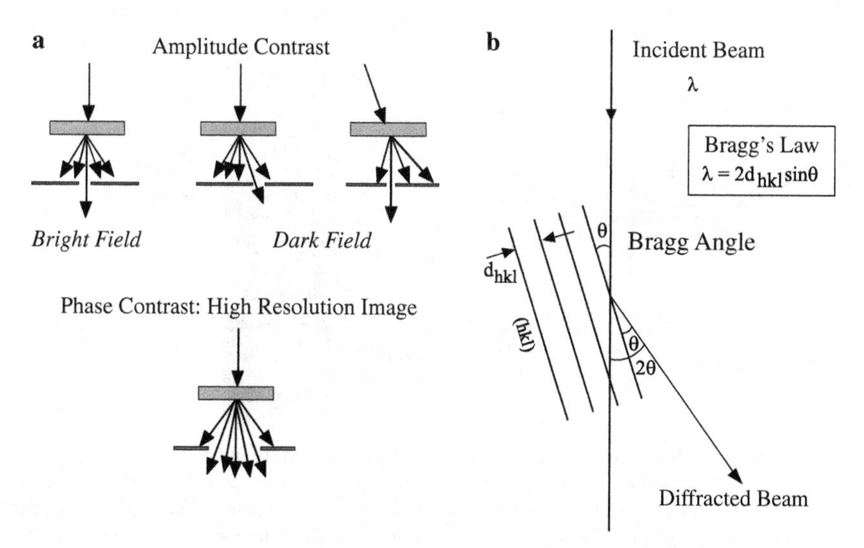

Fig. 3.11 (**a**) Both types of contrast in microscopy: amplitude contrast and phase contrast. (**b**) Bragg's law: diffraction conditions of a crystal lattice plane by an electron beam

scattered by the object recombine to form the image and the resulting image contrast is very weak. Contrast is therefore directly related to variations in the number of electrons scattered or diffracted by the sample.

Phase contrast has another origin: This involves the interference of several diffracted beams with the incident beam in order to form the image.

As described in the preceding case, contrast is very weak or even nonexistent (especially for a very thin sample). Yet the diffracted beams (at least in the approximate framework of the weak phase) have undergone a phase shift of $\pi/2$ while passing through the sample. Any objective lens defocus can produce an additional phase shift of $\pi/2$, resulting in interferences in the image plane between diffracted beams and the incident beam. These interferences correspond to the contrast in high-resolution electron microscopy (HRTEM).

5.2 Diffraction Contrast Imaging Modes in TEM and TEM/STEM

The dual role of the objective lens, using a set of magnification lenses, is to simultaneously enable observation of the image plane and the back focal plane that contains the specimen's diffraction pattern. The combination of diffracted beams coming from the object – focused by the objective lens – forms the magnified image of the object in the Gaussian plane, as well as its diffraction pattern in the Abbe plane. Depending on the size and position of the objective apertures located within this plane, phase contrasts (HRTEM) or amplitude contrasts (bright field, dark field, and weak beam) can be made as a function of the scattered or diffracted beams selected (Fig. 3.12).

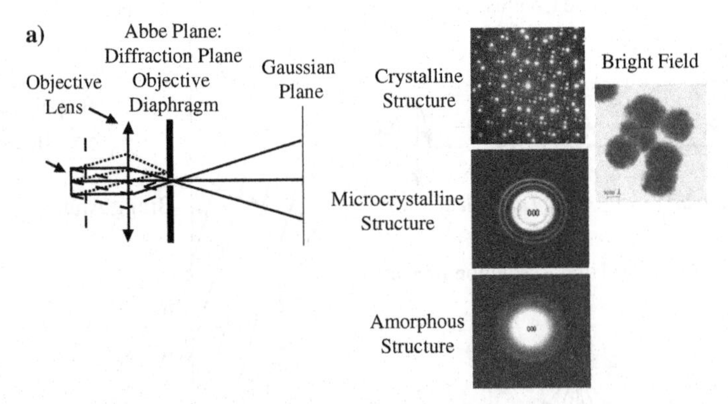

Fig. 3.12 Different diffraction contrasts in the imaging modes: (**a**) bright-field and (**b**) dark-field imaging modes. In the diffraction plane, three examples of diffraction patterns are shown: a crystalline structure, a microcrystalline structure, and an amorphous structure. The bright-field image (**a**) situated in the image plane of the objective lens corresponds to monocrystalline spherules, for which the objective aperture is centered on the optic axis. The dark-field image (**b**) is taken with an objective aperture centered on a portion of the diffraction rings

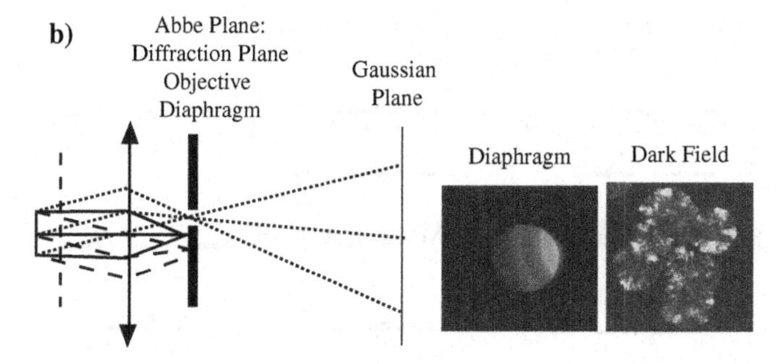

Fig. 3.12 (continued)

The morphology of a material will be observed in contrasted bright-field mode, with contrast due to crystallographic orientation in dark-field mode and with atomic structure contrast in HRTEM mode.

Using a set of various selected area apertures (located in the Gaussian plane), the size of the diffracting area in SAED mode can be decreased down to approximately 100 nm. Below this size, the probe size will directly limit the area to be analyzed. This area can be as small as 2 nm in microdiffraction, nanodiffraction, CBED, LACBED, and nanoprobe mode. Diffraction-contrast imaging can therefore be performed using a static parallel and/or convergent beam.

5.3 Chemical Contrast Imaging Modes in TEM and TEM/STEM

Chemical imaging in a convergent-beam electron microscope or TEM/STEM can be made using X-rays emitted by the target and inelastic transmitted electrons' energy losses. As the X-rays' emission yield increases with the atomic number (Z), the chemical contrast of light elements will be more difficult to obtain. This difficulty will be compensated for through the use of energy loss images which are more sensitive to low atomic numbers.

Chemical imaging is performed using energy-dispersive spectrometry (EDS) of the X-rays emitted by the target. It is used to simultaneously recover all characteristic X-rays of chemical elements using an X-ray detector consisting of a diode made of a doped semiconductor (Si or Ge) that is located above the sample. This detector is placed inside the objective lens at an optimal angle based on the TEM sample holder, close to the sample in order to optimize signal yield.

Electron energy-loss imaging can be performed using an electron energy-loss spectrometer located below the viewing screen, or through a prism or Omega energy-loss filter situated in the column. The first type of imaging involves the use of a convergent beam in STEM mode, i.e., a convergent beam scanning the surface of the specimen. In the second case, imaging is made directly on a conventional TEM in a static parallel-beam mode.

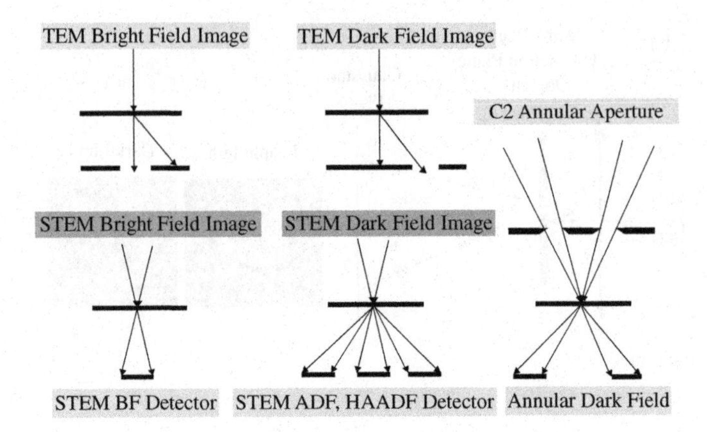

Fig. 3.13 Different microscopic imaging modes: parallel beam and convergent beam

As long as the incident beam is static parallel or convergent scanning, we can make bright-field and dark-field images in TEM or STEM mode (Fig. 3.13). In STEM mode and annular dark-field mode, we can produce chemical images whose resolution reaches the atomic level when the size of the probe is smaller than the interatomic distance: annular dark-field (ADF) and high-angle annular dark-field (HAADF) modes. The introduction of an annular aperture at the C2 condenser level (Fig. 3.13) or a triangular aperture at the level of the objective diffraction plane also enables imaging in atomic-number contrast mode (Z-contrast).

5.4 Spectroscopic Contrast Imaging Modes in TEM and TEM/STEM

Transmitted electron energy-loss spectroscopy (EELS) is used to characterize chemical elements. The threshold position indicates the depth of the excited level with regard to the Fermi level. The peak amplitude is characteristic of the quantity of atoms involved in this signal. The energy-loss threshold structure enables the determination of the chemical bond type for the element analyzed and the study of fine structures that follow the threshold (EXELFS), enabling the calculation of the number of this element's neighbors. In energy-filtered imaging mode (EFTEM), bright-field images can be obtained from the zero-loss peak using the elastic signal only; dark-field images can be obtained from inelastic peaks specific to the atoms of the target.

There are two types of spectrometers (Fig. 3.14a, b) available for spectroscopic analysis, enabling the parallel collection of elastic and inelastic electrons to perform the energy-filtered TEM imaging (EFTEM) and EELS spectroscopy. The first type, the one most commonly used, is an electron energy-loss spectrometer (PEELS) located outside the column, below the microscope's viewing screen. The second

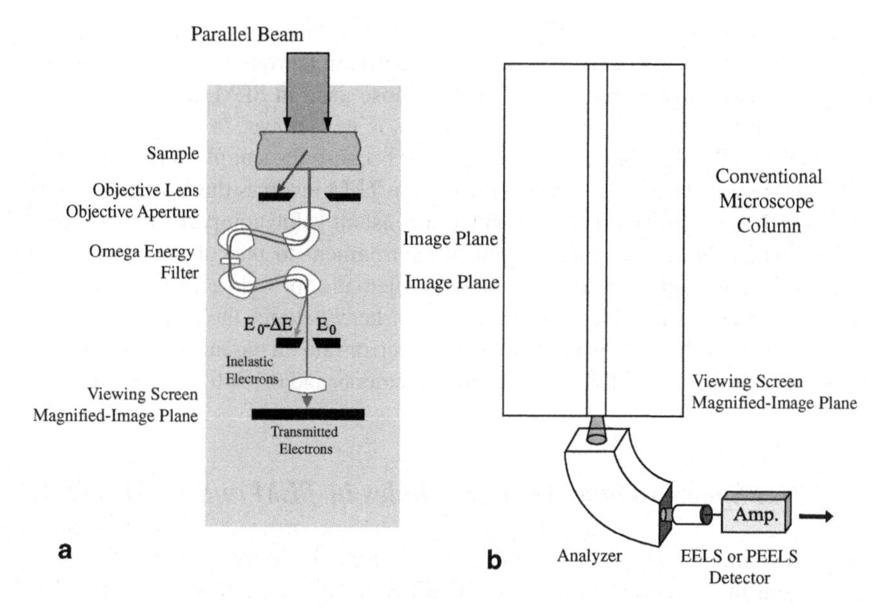

Fig. 3.14 Different types of energy filters: (**a**) an Omega filter located in the column and (**b**) a PEELS filter located outside the TEM column

type is an Omega filter situated in the column. PEELS imaging can be made in parallel beam, fixed convergent beam, or scanning mode. The important difference between these two types of detection is the sample surface analyzed. The first mode is often used to analyze interfaces up to nanometric scales; the image observed on the screen is very small (1 cm^2). On the other hand, EFTEM imaging with an Omega filter is used to observe a large sample area (i.e., the whole surface of the viewing screen), and because of this feature, this tool is frequently used in biology.

5.5 EDS Chemical Analysis Methods in TEM and TEM/STEM

Chemical analysis can be performed using an EDS (energy-dispersive spectroscometry) detector, which uses a fixed beam or a scanning beam as in SEM or STEM mode with a probe size down to 1 or 2 nm, depending on the microscope. The focused beam can induce sample contamination and cause radiation damage that may result in material decomposition or destruction. To prevent or minimize these problems, which vary depending on the chemical nature of the material, it is necessary to use a cooling stage with a liquid-nitrogen-cooled sample holder.

An EDS detector is used to simultaneously acquire the analytical spectrum of all the chemical elements of a material, starting with the light elements, on a multichannel screen.

Quantitative or semi-quantitative element analysis requires the use of a known external standard or an internal standard within the sample. The problems related to quantitative EDS analysis in TEM are different from those of the SEM because of sample thickness (5–200 nm). The same ZAF correction programs (atomic number, absorption, and fluorescence corrections) as those used in SEM and WDS microprobe cannot be used, as the sample thickness is not known. In the programs for TEM analysis, fluorescence effects are ignored because the sample is thin and there is little fluorescence. The use of quantification in TEM involves the *K*-factor calibration for each element, i.e., its emission has a constant value compared to a standard element, which has an identical chemical environment to that of the phase to be analyzed. These challenges are especially problematic for light elements. Whereas quantitative analysis provides correct results for heavy atoms, the quantitative analysis of light atoms by EDS is not resolved. The error rate on the molar concentration of chemical elements in EDS is 5% for light elements and 0.1% for heavy elements.

5.6 EELS Spectroscopic Analysis Modes in TEM and TEM/STEM

Energy-loss analysis is used to quantitatively measure the concentration of a chemical element in a material. In the case of EELS quantitative analyses, the problem of standards is even more pointed since this technique is used to determine the type of chemical bond as well as the chemical environment of each element. In addition to this aspect is the fact that this method provides the best results for the analysis of light elements, making it a complementary technique to EDS. The error rate on the molar concentration of chemical elements is 0.1% for light elements and 5% for heavy elements.

6 Conclusion and Information Assessment

A specimen can be analyzed and studied in rather different ways if there are detectors for each signal emitted by the sample during its interaction with an electron beam. Depending on the signal used, the information is quite different:

- *Secondary electrons* are used to observe surface topography, which can be performed using an SEM or a TEM.
- *Backscattered electrons* are used to produce images of phase distribution and surface topography for thick samples in SEM. They will also help to highlight crystallographic orientations using the electron backscattered diffraction mode (EBSD).
- *X-rays* are used to perform qualitative or quantitative elemental analysis as well as chemical mapping. These analyses can be performed in SEM and TEM.
- *Auger electrons* are used to identify the chemical species of the superficial layers of the sample (1 nm maximum), which can be done using an SEM dedicated to spectroscopy.

Table 3.2 Characteristics of responses to the electron–matter interaction, information on the image obtained and the means used, as well as the spatial resolution of the images

Emission	Energy	Characteristics	Image information	Means	Spatial resolution
Secondary electrons	<50 eV	Emitted by the surface <50 nm	Morphology–Topography of the sample surface	SEM TEM/STEM	1–5 nm
Backscattered electrons	Energy	Output = $f(Z^3)$	Z-contrast images	SEM	
	100 keV to few hundred eV	$5 < D < 200$ nm	Supplement to microanalysis	TEM/STEM	< 10 nm
Auger electrons	Intermediate < 2 keV	Emission from first atomic layers	Chemical analysis of the surface	SEM	1 μm
		Information on the crystallographic structure	Morphology–microtexture defects diffraction	TEM	0.07–0.2 nm
Transmitted electrons	$E_0 - \Delta E$		Chemical analysis	TEM/STEM	Minimal area analyzed 0.5 nm
		Electronic structure	Chemical bond and chemical environment	STEM EELS	<0.5 nm
			Quantitative	SEM	1 μm
X photons	Few keV	Information on the nature of the elements	Elemental analysis of the target	TEM/STEM	Minimal area 0.5 nm
Cathodo-luminescence UV, IR, and visible	Few keV	Electron–hole pair recombination	Highlighting impurities and defects	SEM	1 μm

- *Electrons transmitted* by elastic scattering are used in TEM or TEM/STEM to identify the microstructure through the combination of conventional modes: bright-field, dark-field, weak-beam, microdiffraction, lattice fringe, and high-resolution imaging modes. They will be used to obtain structural imaging, crystallography, chemical and spectroscopic imaging, and chemical and spectroscopic data.
- *Electrons transmitted* by inelastic scattering will be used to identify the chemical nature of the elements, the nature of chemical bonds, and the atomic environment (EELS in a TEM, TEM/STEM, or STEM).

Table 3.2 lists the different responses to the electron–matter interaction, the characteristics of these signals, the image information, the equipment used, as well as the resolution of these images.

For bulk samples, the SEM will make it possible to study morphology with good resolution (1 nm), as well as to perform semi-quantitative or quantitative elemental analysis on the scale of a few nanometers. The latest-generation SEMs produce even better results, as they are equipped with FEG electron sources.

If good resolution power is desired, then it is necessary to limit the excited volume. To do this, it is necessary to work on *thin samples* in TEM, TEM/STEM, or STEM. The goal is to be able to fully characterize the specimen, i.e., to precisely determine the texture and the local microstructure, as well as to be able to determine the local chemistry of particles a few nanometers in size.

Bibliography

Ammou, M. (1989). *Microcaractérisation des solides. Méthodes d'observation et d'analyse*, CRAM CNRS S. Antipolis.

Angenault, J. (2001). *Symétrie et structure cristallochimique du solide*. Vuibert, Paris.

Ayache, J. and Morniroli, J.-P. (2001). *Microscopie des défauts cristallins* (ed. Société Française des Microscopies). École d'Oléron, Paris.

Barna, A., Radnoczi, G., and Pecz, B. (1997). Preparation techniques for electron microscopy. In *Handbook of Microscopy Application in Material Science*. VCH, Weinheim.

Bethge, H. and Heydenreich, J. (1987). *Electron Microscopy in Solid State Physics*. Elsevier, Amsterdam.

Colliex, C. (2004). *La microscopie électronique*. Que sais-je – PUF.

Delain, E., Fourcade, A., Révet, B., and Mory, C. (1992). *Microsc. Microanal. Microstruct.*, **3**, 175.

Delain, E. and Le Cam, E. (1995). The spreading of nucleic acids. In *Visualization of Nucleic Acids* (ed. G. Morel). CRC Press, Boca Raton, London, Tokyo.

Eberhart, J.-P. (1989). *Méthodes Physiques d'étude des minéraux et des matériaux solides*. Dunod BORDAS, Paris.

Georges, J.-M. (2000). *Frottement, usure et lubrification, Sciences et techniques de l'ingénieur*. Eyrolles, Paris.

Goldstein, J.I., Newbury, D.E., Echlin, D.C., Romig, A.D., Lyman, C.E., Fiori, C., and Lifshin, E. (2003). *Scanning Electron Microscopy and X-Ray Microanalysis*, 3rd edition. Kluwer Academic/Plenum Publishers, New York.

Goodhew, P.J. (1985). Thin foil preparation for electron microscopy. In *Practical Methods in Electron Microscopy*, vol. II. Elsevier, Amsterdam.

Hawkes, P. (1995). *Electrons et microscopes – vers les nanosciences*. CNRS Editions, Belin.
Hirsch, P.B., Howie, A., Nichols, R., Pashley, D.W., and Whelan, M.J. (1977). *Electron Microscopy of Thin Crystals*, vol. 13. R.E. Krieger, New York.
Jensen, P. (2001). *Entrée en matière: les atomes expliquent le monde?* Seuil, Paris.
Jouffrey, B., Bourret, A., and Colliex, C. (1983). *Cours de l'école de microscopie électronique en science des matériaux, Bombannes*. CNRS, Paris.
Marioge, J.-P. (2000). *Surfaces Optiques. EDP Sciences*, 229–231.
Maurice, F., Meny, L., and Tixier, R. (1987). *Microanalyse, microscopie électronique à balayage*, Ecole d'été 1978. Les éditions de Physique.
Morel, G. (1995). *Visualization of Nucleic Acids*. CRC Press, Boca Raton, London, Tokyo.
Morniroli, J.-P. (2002). Large-angle convergent-beam electron diffraction. In *Applications to Crystal Defects* (ed. Société Française des Microscopies). Monograph of the French Society of Microscopies, Paris.
Newbury, D.E., Echlin, P., Fiori, C.E., Joy, D.C., and Goldstein, J. (1986). *Advanced Scanning Electron Microscopy and X-Ray Microanalysis*. Plenum Press, New York.
Ratner, M. and Ratner, D. (2003). *Nanotechnologies: La révolution de demain*. Campus Press, France.
Sherzer, O. (1949), *The theoretical resolution limit in the microscope*, JAP 20, 20.
Spence, J.C.H. and Zuo, J.M. (1992). *Electron Microdiffraction*. Plenum Press, New York and London.
Wautelet, M. (2003). *Les nanotechnologies*. Dunod, Paris.
Willaime, C. (1987). *Initiation à la microscopie électronique à transmission*. Société Française de Minéralogie et Cristallographie.
Williams, D.B. (1984). Practical analytical electron microscopy. In *Materials Science* (ed. Philips Electronic Instruments, Inc.). Electron Optics Publishing Group, New Jersey.
Williams, D. and Carter, B. (1996). *Transmission Electron Microscopy*. Plenum Press, New York.

Chapter 4
Materials Problems and Approaches for TEM and TEM/STEM Analyses

1 Introduction

Characterizing a material's microstructure comes down to determining the morphological, textural, structural, and chemical parameters of this material. To respond to a problem presented by a given material, it is necessary to define the pertinent scale for investigating its microstructure. Before beginning a microstructure investigation using transmission electron microscopy, one must first determine the results obtained at a larger scale by using several types of analyses at different locations in the material using different spatial resolutions.

To do this, structural analyses resulting from X-ray and electron diffraction can be compared. Chemical analyses can be performed by characterizing the X-rays coming from the sample using energy-dispersive spectrometry (EDS), wavelength-dispersive spectroscopy (WDS), or secondary-ion mass spectrometry (SIMS). Other spectroscopic analyses are based on the interaction of photons with the material (X-rays photoemission spectroscopy (XPS), Rutherford backscattering spectroscopy (RBS), nuclear reaction analysis (NRA), particle-induced X-ray emission (PIXE), Raman spectroscopy) or on the investigation of electrons resulting from the interaction of the electron beam with the sample electron spectroscopy for chemical analysis (ESCA), Auger electron spectroscopy (AES), and electron energy loss spectroscopy (EELS). Table 3.2 shows the different spatial resolutions of analyses using electron microscopy.

2 Analyses Conducted Prior to TEM Analyses

The size of the domains to be characterized, their concentration, and their distribution will determine the spatial resolution required for observations. This will result in the use of one technique over another. A detailed microstructural investigation must include different characterization steps, from the macroscopic scale through the mesoscopic and microscopic scales, down to the nanoscopic scale. All of the characterizations and their scales of observation are listed in Table 4.1.

J. Ayache et al., *Sample Preparation Handbook for Transmission Electron Microscopy*,
DOI 10.1007/978-0-387-98182-6_4,

Table 4.1 Scales of characterization: macroscopic, mesoscopic, microscopic, and nanoscopic

				Type of analysis				
	Analyzable surface area	Spatial resolution	2D–3D observation	Topography	Morphology	Structure/orientation	Chemical analysis	Properties
X-rays	cm	1 μm	3D reconstruction			+++	+++	
Photonic	cm	400 nm	2D	+++	++	+		+++
Confocal		200 nm	3D reconstruction		+++			+++
SEM	cm	50–1 nm	3D	+++	+++	+	+++	++
TEM	mm	0.1 nm	2D	+	+++	+++	+++	+++
		2 nm	3D reconstruction	++	++	+++		+++
AFM	mm	0.2–10 nm	2D (ab) plane	+++	+++			++
		0.1–1 nm	3D (z)	+++	+++			++

2.1 Macroscopic Characterization

Structural analyses begin at the macroscopic scale. This corresponds to what can be observed with the naked eye alone or by using a magnifying lens. This type of analysis helps to determine whether the material is bulk, porous, or in liquid phase; homogeneous or heterogeneous; either mixed or with large grains.

2.2 Microscopic Characterization

Structural analyses at these scales can be carried out using optical and confocal microscopy, as well as by scanning or transmission electron microscopy. They may be combined with X-ray imaging and SIMS imaging.

Photon characterization (down to 400 nm) is used to identify morphology, e.g., the size of grains composing the material, their orientation, or their texture. Classic optical microscopy techniques are used to observe the different phases. Crystalline materials have a strong phase contrast related to their high scattering power (high atomic number or Z), and they can be directly observed. In the case of a poorly diffusive material such as organic or biological materials, preliminary staining often precedes observation in order to increase the optical density of phases to be observed.

At these scales, observations under polarized light can clearly show the presence of grains (from 1 μm) and their various orientations and can also highlight a preferential orientation. The polarizing microscope is equipped with a polarizer in the lighting system, which selects a polarization orientation, and a second rotating polarizer (called an analyzer), which selects the light rays once again based on another polarization direction. When the polarizer and the analyzer are oriented at a 90° angle, which corresponds to phase-contrast imaging conditions (also called "crossed Nichols" conditions), the crystal planes selectively diffuse the polarized light depending on their orientation. Under these conditions, an amorphous material that has uniform contrast (i.e., the same polarization color in all directions) can be distinguished from a crystalline or polycrystalline material, which changes color depending on its orientation to the light (Fig. 4.1).

Confocal microscopy uses a highly convergent laser beam to produce images with very little depth of field. Usually the laser beam scans the sample surface, helping to improve resolution. By focusing the objective at different depth levels in the sample, a series of images can be produced, from which a three-dimensional representation of the specimen can be made. This can be used to make virtual cross sections of the sample. Coupled with fluorescent labeling, this type of observation allows for the in situ viewing of dynamic phenomena in live tissues.

For materials in solid-state physics, the overall crystalline structure of the sample must be determined using X-ray diffraction or neutron diffraction techniques. These analyses average the information of the analyzed sample over a volume of 1 μm^3. They provide information on all crystalline phases present in the material, but do

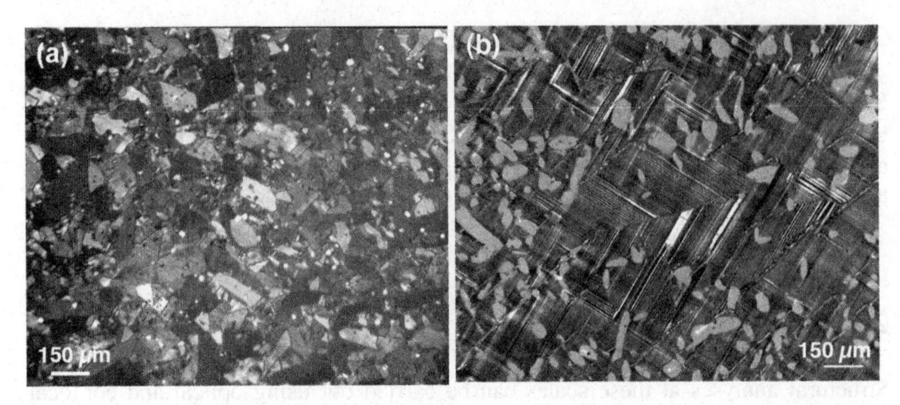

Fig. 4.1 Optical microscopic pictures of a single-phase polycrystalline and a textured ceramic, observed under crossed Nichols polarized light conditions. Each color (represented in *gray scale*) corresponds to a particular orientation. In the image (**a**), a significant number of orientations are present. In the image (**b**), only two perpendicular orientations in the matrix are visible (*J. Ayache, CSNMS-IN2P3-CNRS, Orsay*)

not allow them to be precisely located. These structural analyses have a spatial resolution on the micron scale, comparable to that of the optical microscope. They will be determinant and complementary to structural analyses made at smaller scales.

X-ray imaging is done using X-ray tomography, enabling the 3D reconstruction of the material and highlighting the location of phases.

Ion microscopy is used to obtain an image of the ions present in the sample using a light atom ion beam. During ionic analysis, the sample surface is pulverized by the ion beam. As a result, information can be obtained on the ionic concentration profile of a given chemical element through the depth of the material.

This type of analysis is suitable for thin film layers or multilayer materials. The spatial resolution of these analyses, on the order of a micron, is close to that of X-ray analyses.

2.3 Microscopic and Nanoscopic Characterization

The morphological, structural, and chemical characterization at this scale can be carried out using scanning electron microscopy, transmission electron microscopy, and atomic force microscopy. At this scale, electronic tomography picks up where X-ray tomography left off, since its spatial resolution is that of transmission electron microscopy. It is also used to reconstruct small volumes whose dimensions are on the order of magnitude of the microstructural defects. Complementary chemical and spectrometric analyses can be carried out with techniques using the dispersion of X-rays (EDS or WDS) or electrons (ESCA, AES) and, in some cases, using SIMS analysis.

What must be kept in mind is that because of the various imaging and diffraction modes, transmission electron microscopy is the domain of structural characterization of multiphase materials whose phase size can range from the micron to the sub-nanometer. TEM analyses can be carried out down to the atomic and molecular scales. Table 4.2 shows the scales of the various analyses.

Table 4.2 Scales of the various TEM analyses

Macro *Meso* *Micro* *Nano*

1 mm *100 µm* *10 µm* *1 µm* *100 nm* *10 nm* *0,1 nm*

Topography

Structure, Crystal Defects

Crystallography, Diffraction

EDS Analysis and Imaging

EELS Analysis and Imaging

TEM *HREM*

HREM *HAADF*

3 Approach for Beginning the Investigation of a Material

Microstructure identification and characterization are indispensable to understanding the elemental mechanisms governing, for example, the behavior of a material under various stresses. Understanding the structure–properties relations generally requires knowledge of the material's structural components and/or its defects at all scales, ranging from dimensions down to the atomic scale (Table 4.2). In practice, depending on the problem to be dealt with, a narrower range of observation scales may suffice, but it should never be reduced to just a single dimension. Therefore, the first step is to precisely define the problem and the level of observation scale necessary for characterization. The investigation of a material using transmission electron microscopy involves bringing together bibliographic data combining the material's overall characteristics. It is helpful to know the material's history (processing, growth mechanism, environment of use, aging, etc.). Finally, the results of other analyses done beforehand must be combined. Once all the data are brought together, characterization using transmission electron microscopy can begin. Table 4.3 shows one recommended approach.

Regardless of the observation or analysis techniques used to approach a specific materials problem, the interpretation of the results must take some important

Table 4.3 Different observation scales to be used before beginning preparation of the TEM sample

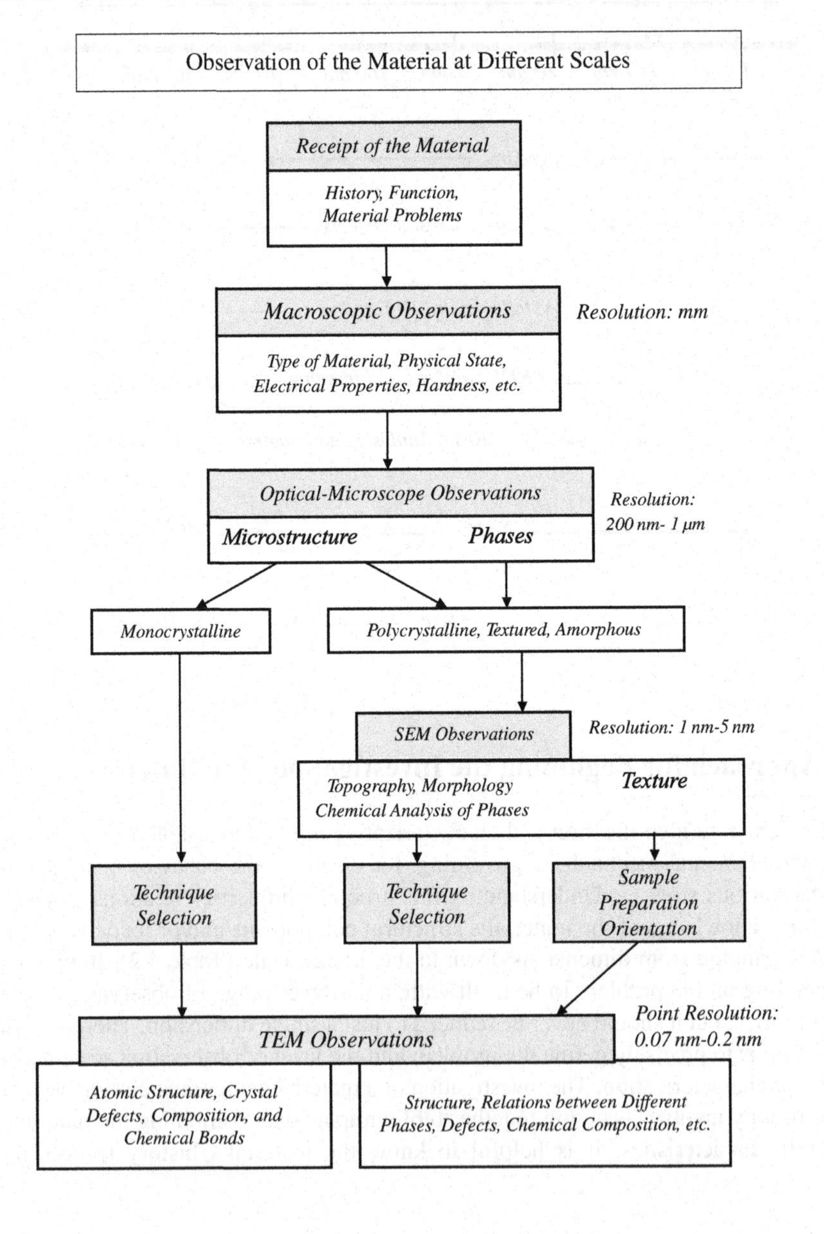

parameters into account, e.g., the reliability and precision of the analyses as well as the volumes analyzed. Any technique has risks of error that can usually be circumvented by comparing the results of various techniques.

In cases of doubt, combining results will help to confirm a microstructure or a structure at the atomic or molecular scale. This comparison of results will then lead to an understanding of the physical properties associated with the microstructures, as well as their behavior under different stresses.

4 Selection of the Type of TEM Analysis

The selection of techniques to be used for TEM characterization of a material depends on the material problem with regard to its structure, properties, and application. This choice depends on several criteria:

- *Type of investigation:* surface or volume investigation.
- *Type of macroscopic organization of the material:* bulk, single-layer, multilayer, or fine particle material.
- *Number of chemical phases:* single-phase or multiphase, the phases to be characterized, their dimensions, as well as their location and distribution.
- *Type of crystalline or amorphous organization of the microstructure:* amorphous, poorly organized, nanocrystalline, microcrystalline, polycrystalline, or monocrystalline.
- *Type of defects to be analyzed:* point defects (0D), extended crystal defects (1D, 2D, and 3D), twins, grain boundaries, interfaces, dislocations, segregations, precipitates, etc.
- *Type of physical, chemical, or functional properties to be investigated in relation to the structure:* optical, electrical, electronic, magnetic, mechanical, chemical, or functional (location of functional sites).

Lastly, it is necessary to take into account the resolution required for observing the microstructural defects or components. This is based on the dimensions of the microstructure, which can vary from a millimeter (inclusion) to a fraction of a nanometer (point defect). Sometimes, this results in a dual situation. It is necessary to account for both the dimensions of the defect, requiring one scale of observation, and the distribution and density of these defects responsible for its properties, requiring another scale.

5 Analysis of Topography

Generally speaking, surface-morphology depend on the growth mechanism and can have various types of topographies (Fig. 4.2). Their analyses are carried out using

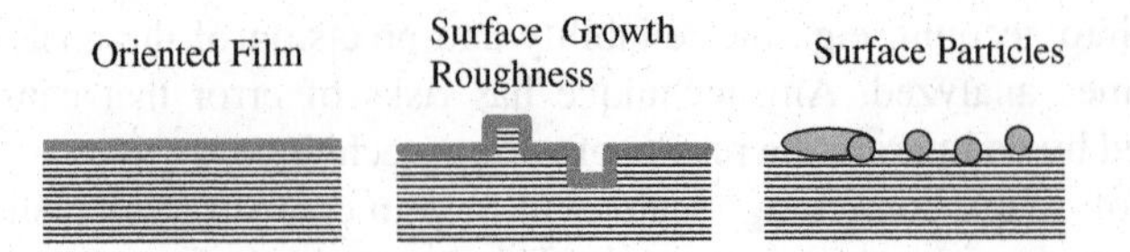

Fig. 4.2 Different types of surface topographies

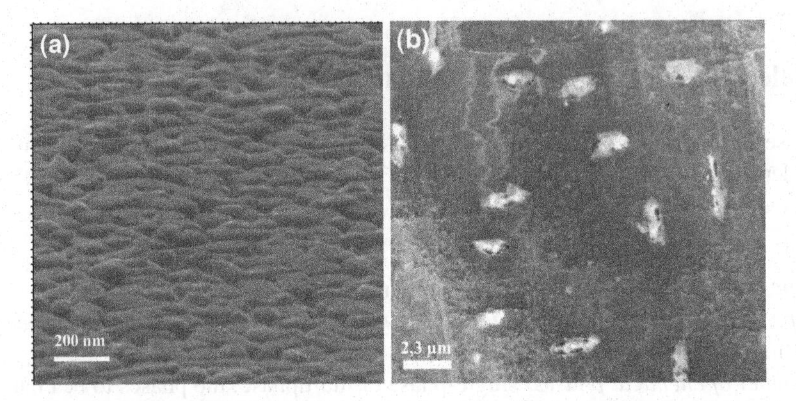

Fig. 4.3 (**a**) AFM picture of the surface of a PZT ferroelectric film deposited by laser ablation. (**b**) TEM plan-view picture showing surface precipitates in a superconductor film of $YBa_2Cu_3O_7$ (*J. Ayache, CSNM-IN2P-CNRS 1341, Orsay*)

SEM and AFM (Fig. 4.3a) microscopies. However, the structural analysis of a surface film deposited on a substrate, a thin film, or a material dispersed upon a surface can also be performed using a TEM which brings, in addition, structural information (Fig. 4.3b).

6 Structural Analysis in TEM

6.1 Morphology and Structure of Materials

These are investigations related to the microstructure of the material volume.

Morphological analyses of volume phases can be conducted on all types of materials and macroscopic organizations in which the material is found: bulk, single-layer, multilayer, or fine particle materials (Fig. 4.4). Structural analyses help to identify the number of chemical phases present in bulk material depending on whether it is single phase or multiphase.

Generally for materials in solid-state physics, structural analysis may be carried out by combining observation modes, depending on the sample type: bright field, dark field, filtered imaging, diffraction, chemical analysis, and spectroscopic analysis. Morphology observation is mainly performed in imaging mode. It also includes statistical analysis of the shape, size, and distribution of a given component (elongated grains in the example of Fig. 4.3a, b), which can be correlated with morphological analysis.

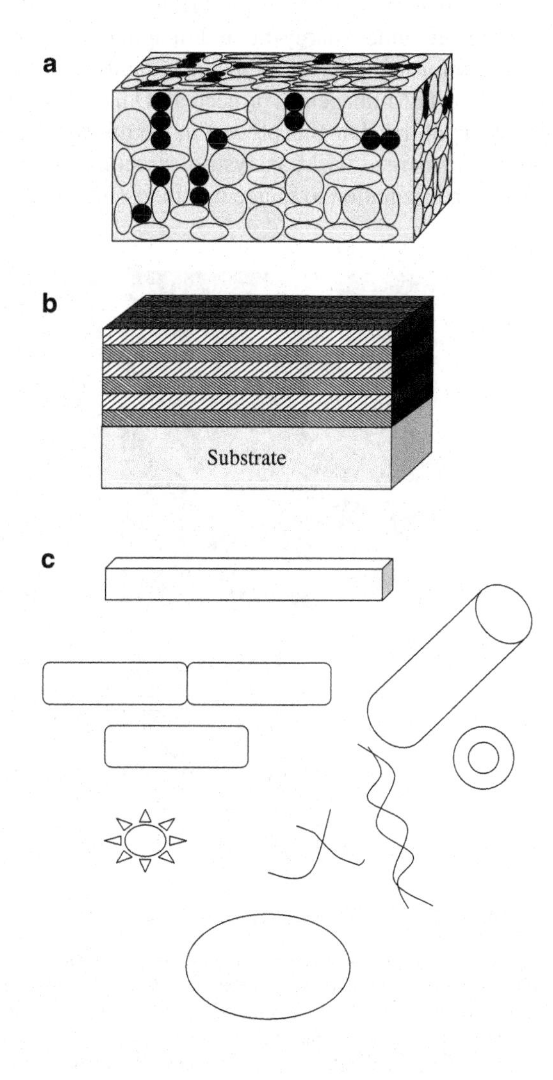

Fig. 4.4 Diagrams showing the three organization types of the material: (**a**) bulk, (**b**) multilayer, and (**c**) fine particles

For mixed–composite and multilayer materials, the investigation is essentially structural and analytical in order to determine each of the constituents and its relationship with neighboring constituents: distribution, layer thickness, and junction quality.

The structural investigation of polymer materials focuses on multiphase polymers and makes it possible to determine the different phases and their coherence. For semi-crystalline polymers, highly localized crystallographic investigations can be planned.

The structure of stabilized bulk biological materials, prepared by ultramicrotomy, is essentially observed in bright-field imaging mode. In certain cases, it can be observed under energy-filtered imaging mode or in annular dark-field mode.

The information given in the image must be conceptualized in terms of volume, because a TEM image is a projected 2D image of a sample. The image is of a plane in which all the elements present in the thickness of the section are found superimposed and merged. TEM analysis of structural components as collagen fibers require specimen orientation to image the fibers (Fig. 4.5).

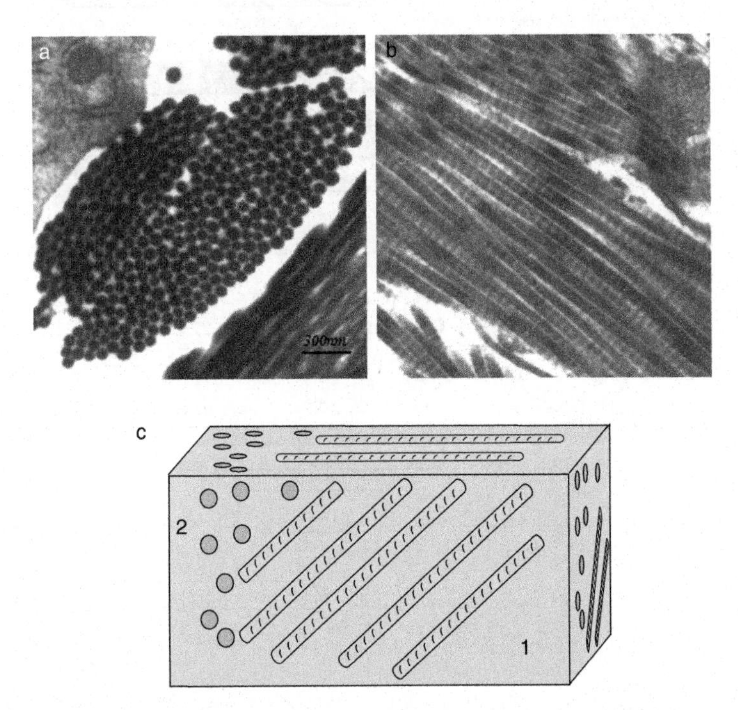

Fig. 4.5 TEM images of collagen fibers: (**a**) cross section and (**b**) longitudinal section. In the latter, we can see the transverse lines of the fiber, which have a periodicity of 65 nm. We may note at the bottom right-hand side in (**a**), collagen fibers viewed from an ordinary longitudinal-plane section do not allow us to determine their diameter and their periodicity. In (**c**), the diagram represents two types of sections in TEM images: (1) longitudinal and (2) cross section (*J. Boumendil CMEABG, Université Claude Bernard-Lyon 1 Villeurbanne*)

In the example shown in Fig. 4.6, a retrovirus can be confirmed as coming out of the cell only if it can be seen budding (arrow in Fig. 4.6a). In the other cases, a conclusion cannot be reached. As Fig. 4.6b shows, superimposing the various planes

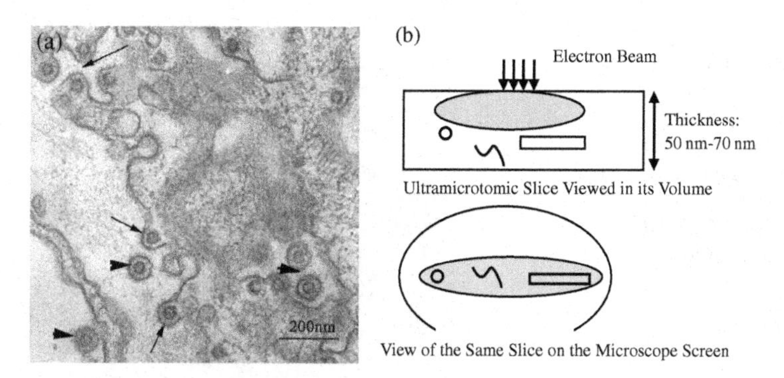

Fig. 4.6 (**a**) TEM bright-field image of a thin section of retrovirus particles on a cell. (**b**) Diagram of the projection of the section on the image observation plane (*A. Rivoire, EZUS CTμ université Cl. Bernard – Lyonl – Villeurbanne*)

makes it impossible to determine whether they are on the same level as the cell or on a different plane through the thickness of the sample.

The recognition of structures (called ultrastructure by biologists) helps to deduce their biological functions (Figs. 4.7, 4.8). The approach to the phenomena of syntheses or degradation of cellular components will use observations combined with preliminary labeling techniques. Investigations of structural defects and/or contaminants can reveal pathology. The sample preparation technique is of utmost importance with regard to the type of observation, which is usually reduced to bright-field imaging mode.

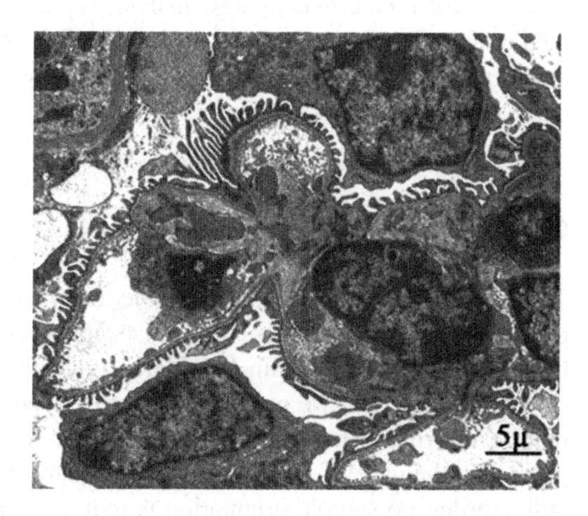

Fig. 4.7 TEM image of a thin section of a renal glomerulus, an organ structure with its different types of cells (*J. Boumendil CMEABG, Université Claude Bernard-Lyon 1 Villeurbanne*)

Statistical analysis of the shape, size, and distribution of a component is sometimes correlated with the morphological analysis. The investigation of serial sections is indispensable to understanding the cell or tissue. With the help of specialized software, their study can produce a 3D reconstruction.

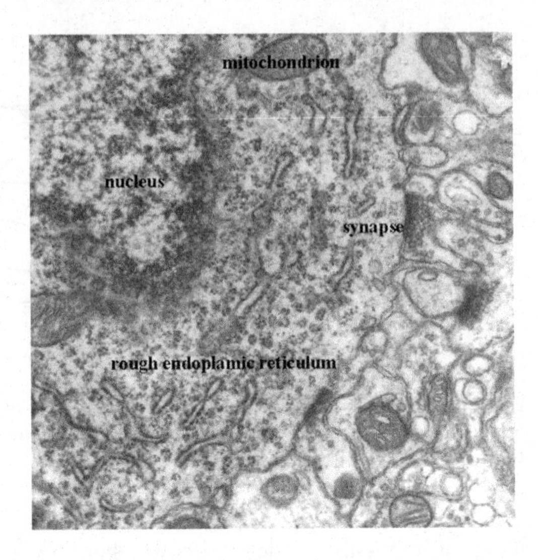

Fig. 4.8 TEM image of a thin section of a neuron, showing its organelles, nucleus, mitochondria, endoplasmic reticulum, ribosomes, and intercellular junctions (synapses) (*J. Boumendil CMEABG, Université Claude Bernard-Lyon 1 Villeurbanne*)

The tomography technique consists of making a series of eucentric TEM images by tilting the sample from +70° to –70° around an axis. This technique is used in conjunction with very powerful software to make 3D reconstructions of cellular organelles.

Textured materials (bone, tooth, collagen, etc.) are considered to be poorly organized or crystalline materials. These types of materials are observed in diffraction contrast.

For isolated macromolecules, analysis is performed in either bright-field or dark-field imaging mode, depending on the case. It is sometimes coupled with molecular modeling to build a 3D reconstruction of these systems. These analyses are generally made in cryomicroscopy (see Section 10.1).

6.2 Atomic Structure

These analyses are generally applicable to crystalline materials as well as to certain poorly organized materials. They also apply to crystal defects such as dislocations, stacking faults, grain boundaries, interfaces, etc. The HRTEM, Z-contrast (HAADF), and STEM modes enable this type of investigation. These analyses require rigorous sample orientation in order to achieve diffraction of the atomic planes. In the case of poorly organized or amorphous materials with a local crystalline order, no sample orientation is required since all orientations are present at the same time in this type of sample. They often only yield lattice fringes (Fig. 4.9).

The atomic structure of phases or grain boundaries and interfaces may be determined by HRTEM. The optimum resolution of HRTEM varies depending on the microscope, accelerating voltage, and spherical aberration of the objective lens. Two types of high-resolution images can be produced, resulting in either an intensity image (Fig. 4.10a) that is obtained directly or a phase image that is calculated from

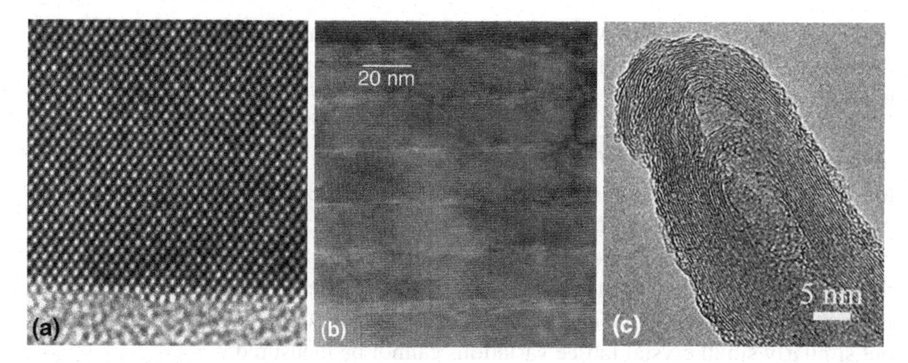

Fig. 4.9 Illustration of atomic planes in (**a**) a bulk perfect silicon single crystal (*J. Ayache, CSNSM-IN2P3-CNRS, Orsay*), (**b**) a $YBa_2Cu_3O_7$–$SrTiO_3$ multilayer material (*J. Ayache, CSNSM-IN2P3-CNRS, Orsay*), and (**c**) lattice fringes in a carbon nanotube (*Qiang F. Fritz Haber Max Planck Institute, Berlin*)

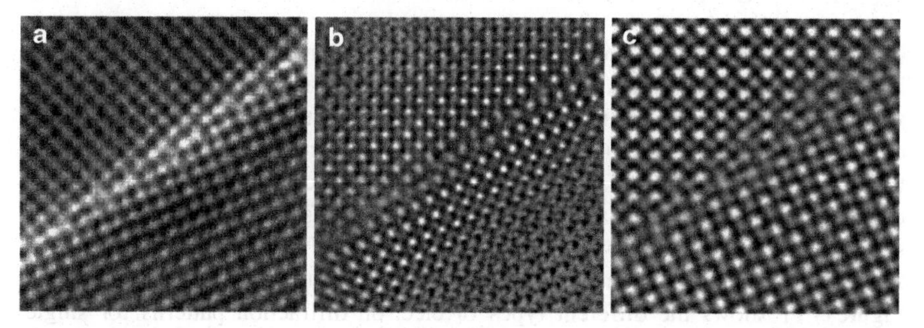

Fig. 4.10 Pictures of a $SrTiO_3$ bicrystal grain boundary produced by three high-resolution techniques: (**a**) HRTEM, (**b**) phase image after reconstructing a focal series of HRTEM images, and (**c**) HAADF chemical contrast image (*J. Ayache, Lawrence Berkeley Laboratory, NCEM, Berkeley, CA, USA*)

the complex wave function of the electrons exiting the sample. This is obtained by reconstructing a focal series of HRTEM images (Fig. 10b). It must be noted that the contrasts obtained are similar, but correspond to different atoms. In Fig. 4.10a, the atomic columns of Sr and TiO are visible; in Fig. 4.10b, the visible columns belong to oxygen and Sr or TiO. The last two columns in Fig. 4.10b are not discernable in this system or under these conditions. The first method serves to unambiguously highlight the arrangement of heavy atoms (which provide a high intensity). The second method allows for better discrimination between light atoms and heavy atoms. The reconstruction of the images leads to the specimen phase and amplitude, i.e., allows it to go back to its crystal potential. The statistical error on the measurements of the crystal lattice parameter is not more than 0.001%. This type of analysis provides quantitative measurements of the crystalline parameters to determine atomic plane relaxation variations at the level of a crystal defect or an interface and gives access to stoichiometry.

HRTEM may be performed with a 200 keV accelerating voltage on samples with a thickness less than 50 nm, with a resolution of 0.23 nm. The resolution can be improved by using higher voltages (300, 400, 800, and 1.2 MeV), allowing for the use of thicker samples. The use of a FEG-equipped microscope enables a limit resolution of 0.07 nm.

Due to a very small probe size of about 0.14 nm (a size on the order of inter-atomic distances), the HAADF technique (or Z-contrast imaging) makes it possible to view the chemical contrast at the atomic scale in TEM/STEM. This contrast is proportional to the square of the atomic number (Z^2). In Fig. 4.10c, the atomic columns correspond to the columns of Sr and TiO; the oxygen columns are not visible. In this mode, the statistical error of the measurement of the crystal parameter is approximately 2%. At the atomic scale, two types of atoms may be differentiated, but similarly small crystal lattice variations cannot be measured.

7 Crystallographic Analysis

Crystallographic analysis consists of determining the orientation of a material and its crystalline structure, i.e., the crystal lattice, symmetry group, and point group, which contains all the symmetry elements of the crystal (position and nature of atoms). Electron diffraction is used to identify a structure known ahead of time through X-ray diffraction (SAED diffraction, nanodiffraction, and microdiffraction) or to determine the crystal lattice and the symmetry class (microdiffraction) or the symmetry class and the point group (CBED). LACBED enables the characterization of crystal defects and quantitative measurement of stresses in a material. Figure 4.11 shows the different types of electron diffraction patterns for single crystal, polycrystal and amorphous structures.

Whereas techniques involving the diffraction of X-rays or neutrons provide information on atomic structure down to the scale of the micron, electron diffraction is used to achieve smaller spatial resolution of the structural analysis at nanometric scales. Convergent-beam electron diffraction also enables the investigation of the structure, the quantification of local parameter variations (near an interface, for example), the characterization of crystal defects (1D, 2D, and 3D), and the localization of defects using large-angle convergent beam mode. Determination of the local structure using CBED is equivalent to that obtained using X-ray diffraction, with the addition of an image of the associated structure. Thus, the diffraction techniques simultaneously associated with the TEM image mode allows the identification of the atomic structure or a defect of the material at the nanometric scale. In some cases using filtered imaging modes, it is possible not only to locate the structural defects at the atomic scale but also to measure the local crystal potential. This will be the challenge in the upcoming years for the new FEG microscopes equipped with C_s aberration correctors, a beam monochromator, and an energy filter (located either inside or outside the TEM column).

Structural analysis using electron diffraction is used to locally identify the structure or structural changes. With electrons, only heavy atoms with a periodic structure are visible. The unambiguous determination of the structure of the sub-lattice con-

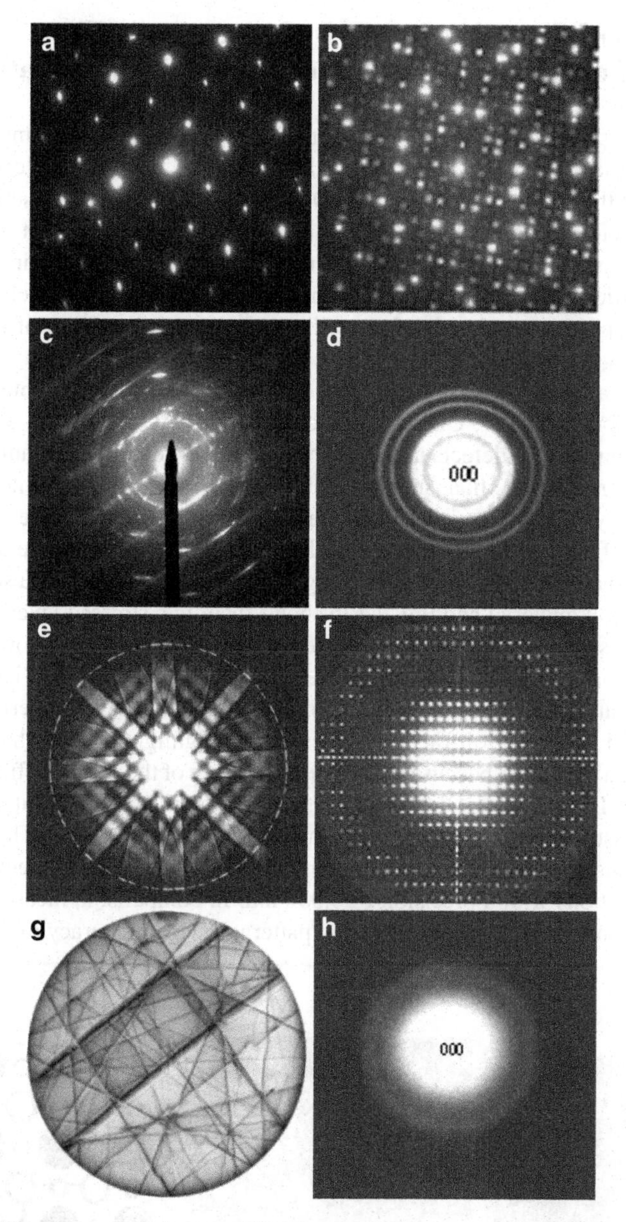

Fig. 4.11 Different types of electron diffraction patterns, from single-crystal structure using SAED (a), microdiffraction (f), CBED (e) and LACBED (g) to bicristal structure (b), polycrystallized (c), microcrystallized (d) and to amorphous structures (h) using SAED mode in SAED (*J. Ayache, CSNSM-IN2P3-CNRS, Orsay*)

taining light atoms, such as oxygen, will require the additional use of neutron diffraction. The major drawback of this technique is the large quantity of material needed (cubic millimeters instead of a few nanometers for the TEM).

8 Analysis of Crystal Defects: 1D (Dislocations), 2D (Grain Boundaries and Interfaces), and 3D (Precipitates)

From an amorphous material to a poorly organized material, then to a microcrystallized material and finally to a monocrystalline material, the extent of crystal order grows first at the atomic scale, then the 2D scale, and lastly to the 3D scale.

The characterization of the resulting microstructures will not result in the same problems. Thus, defects in an amorphous material will be crystalline in nature and the crystal order will be local (a few nanometers). For example, in a well-organized material, a typical example of defects will consist in the formation of amorphous phases at the grain boundaries.

Structural analysis of crystal defects first starts with their orientation using diffraction modes (SAED, CBED, microdiffraction, LACBED). The second step consists of imaging the defects using bright-field, dark-field, weak-beam, HRTEM, or Z-contrast modes. The final step is to subject these defects to a chemical or spectroscopic analysis at the scale of the dimensions of the defect. Figure 4.12 shows two images of twin grain boundaries in the same superconducting ceramic oxide formed due to a phase transformation at high temperature. Figure 4.12a shows them relaxed in a bulk material and Fig. 4.12b shows them in a stressed layer. The effect of growth causes the formation of two types of twins. Figure 4.12c shows this twin wall, which corresponds to an atomic plane common to both domains. The image in Fig. 4.12b is an HRTEM picture of a cross section in a multilayer material.

CBED and LACBED diffraction modes enable highlighting and characterizing the orientation of defects because of the great precision of the Bragg diffraction lines of high-order Laue zone (HOLZ lines). The dynamic diffraction conditions needed for these measurements are obtained in samples thicker than 200 nm. This powerful technique is used to measure crystal parameter variations around defects resulting from deformations with a precision of 0.001 nm. In comparison, measurements of the same parameters on SAED diffraction patterns have an accuracy of 0.1 nm.

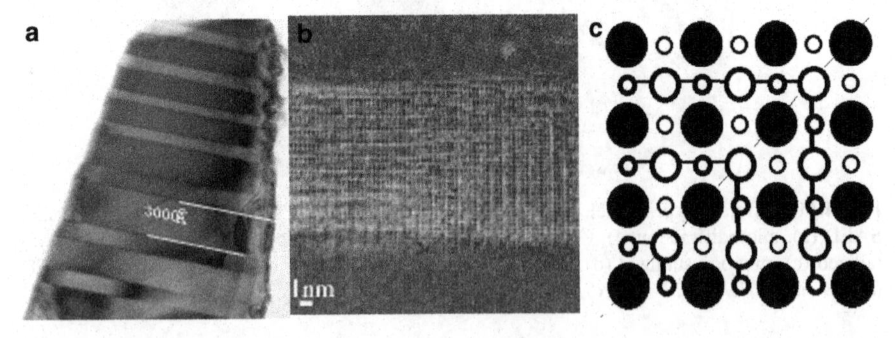

Fig. 4.12 Images of twin grain boundaries in (**a**) a bulk $YBa_2Cu_3O_7$ sample and (**b**) a multilayer $YBa_2Cu_3O_7/SrTiO_3$ sample. (**c**) Diagram of a twin wall showing that the interface is common between the two grains at the atomic scale (*J. Ayache, CSNSM-IN2P3-CNRS, Orsay*)

9 EDS Chemical Analysis and EELS Spectroscopic Analysis

Chemical analysis using EDS and EELS techniques is used to identify the atomic elements of a phase. Spectroscopic analysis is used to determine an atom's chemical bond type, its environment, and the number of close neighbors in the structure.

These analyses may be qualitative and quantitative. In both analysis types, quantitative investigations require standard samples that have the same bonds and chemical environments as the material to be analyzed.

Because the signal of the X-rays increases with the atomic number, EDS analysis will provide a notable signal even for low concentrations of heavy atoms. Maximum sensitivity is 0.2%. The yield from this analysis type depends on the collection angle. The relative error from these analyses varies between 2 and 5%.

EELS analysis can provide a detectable signal only for very low thicknesses (5–50 nm) because of surface plasmons whose energy is low (some 10 eV) and close to the peak without elastic scattering loss (zero loss). The signals emitted at higher energies correspond to direct absorption thresholds, making analysis very sensitive. The lowest detectable concentration is 1%. This technique is a complement to EDS analysis. It provides the best results for the quantitative analysis of light elements.

EDS and EELS analyses can also be used to produce chemical analysis of the elements in a compound (Fig. 4.13).

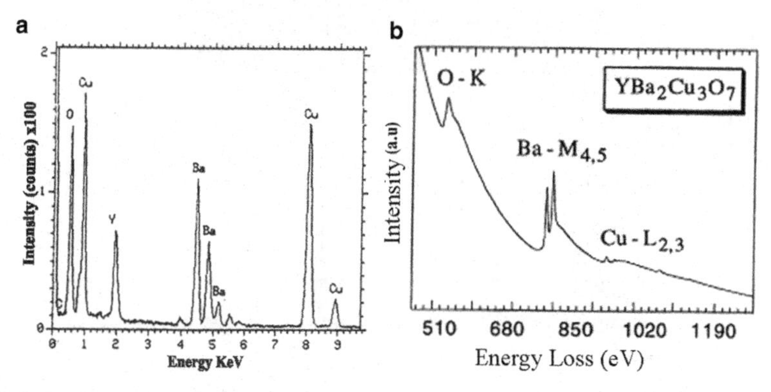

Fig. 4.13 EDS and EELS spectra from a $YBa_2Cu_3O_7$ superconductive oxide (*J. Ayache, CSNSM-IN2P3-CNRS, Orsay*)

9.1 Phase Identification and Distribution

In a multiphase material, chemical mapping is one method for identifying phases and their chemical distribution. It can be conducted using X-rays (EDS) or transmitted inelastic electrons (EELS). These maps can be made in TEM mode in a microscope equipped with an omega filter or in an analytical TEM/STEM or STEM.

9.2 Concentration Profiles and Interface Analysis

Interface analysis requires a vertical orientation to the interface, i.e., one that is parallel to the incident electron beam, in order to analyze only the plane of the interface viewed in image mode. Figure 4.14 shows a bright-field TEM image of interfaces between SiO_2/Ti/Pt/PZT layers deposited on a Si substrate and the concentration profile where the sequence of the multilayer material is found.

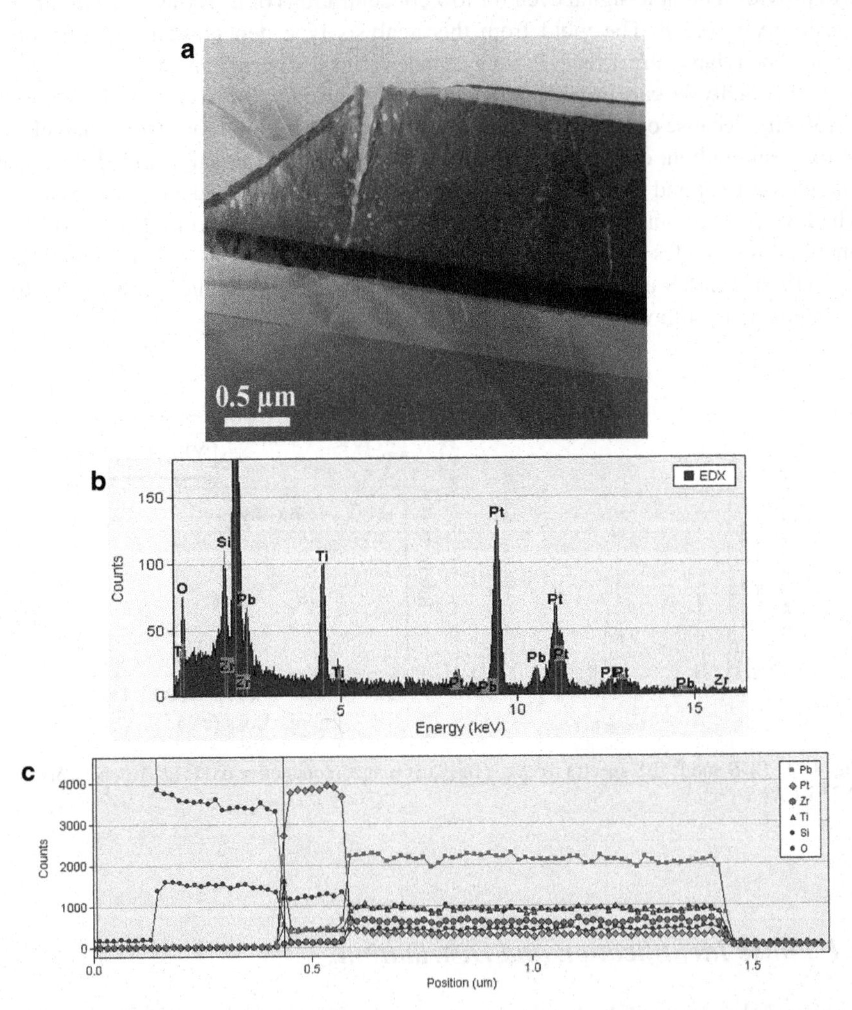

Fig. 4.14 (**a**) Bright-field image of the interface in a multilayer material, SiO_2/Ti/Pt/$PbZrTiO_3$ on a Si substrate, (**b**) point EDS chemical analysis in $PbZrTiO_3$ film, and (**c**) EDS concentration profile across a full interface of $PbZrTiO_3$/substrate (*J. Ayache, Lawrence Berkeley Laboratory, NCEM, Berkeley, CA, USA*)

The spatial resolution required for this type of analysis is on the order of a nanometer and is only obtained in a TEM/STEM microscope equipped with a FEG or a dedicated STEM. In STEM mode, spectrum-image mode can be used, which records both the spectrum and image for the full area analyzed. This results in the formation of chemical and spectroscopic image information along an entire interface.

10 Structural Analyses Under Special Conditions

Two types of analyses can be made: in situ analyses and cryomicroscopic analyses.

10.1 In Situ Analyses

This type of analysis is used to investigate the behavior of a material subjected to different stresses: irradiation, temperature rise or decrease, tension, etc.

10.1.1 At Room Temperature

- *Under irradiation by high-energy electrons* (from 400 to 1,200 keV) to study the dynamics or stability of a system (atomic displacement, phase formation, growth and crystallization, etc.).
- *Under medium- and high-energy ion irradiation* to investigate the dynamics or stability of a system (atomic displacement, phase formation, growth and crystallization, etc.).

10.1.2 At High Temperatures

The in situ handling of samples at high temperatures requires the use of a specific specimen holder. The temperature limit reaches 1,273 K. These experiments require the use of washers or support grids designed for high temperatures.

High temperature in situ experiments also apply to studies under irradiation and those conducted under mechanical stress (investigations of stability, phase transformation, mechanical behavior, growth, etc.).

- *Under irradiation by high-energy electrons* in order to study the dynamics or stability of a system (atomic displacement, phase formation, growth and crystallization, etc.).
- *Under medium- and high-energy electron irradiation* to investigate the dynamics or stability of a system (atomic displacement, phase formation, growth and crystallization, etc.).

- *Mechanical behavior of a material under a stress* (tension or compression) to study the dynamics of dislocation displacement as a function of the stresses applied or already present, as well as to determine dislocation types.

10.1.3 At Low Temperatures

Cryo-observations are made either in a liquid-nitrogen-cooled (77 K) specimen holder or at the temperature of liquid helium (4 K). For superconducting oxides in the family of YBaCuO oxides, whose critical temperature is 92 K, cryo-observation is used to investigate their physical properties in the superconducting state and not at room temperature, where the material is insulating.

The use of a cooled specimen holder also helps to minimize the effects of irradiation under an intense electron beam during observation, especially for local chemical analyses.

10.2 Cryomicroscopy

Cryomicroscopy is a technique developed mainly for biological materials and especially for the investigation of isolated particles such as viruses and macromolecules prepared by the frozen hydrated-film technique or the cryo-ultramicrotomy technique. In addition to the cooled specimen holder, the cryofixed sample itself is cryo-transferred in the microscope in order to be observed under cold conditions. Breaking the cold chain during these transfers must be avoided at all costs, since it leads to ice formation and sample deterioration.

10.2.1 Structure of Isolated Particles from Biological Materials or Polymers

Observations under cryomicroscopy have an advantage over other techniques in that they can be used to view the sample in its native environment without any staining. This type of observation allows for the 3D reconstruction of macromolecules.

In the case of biological macromolecules, they can be observed directly on frozen suspensions in bright-field imaging mode only under "low-dose" conditions (low irradiation of maximum dose by an electron per unit of surface (dose: 1 $e^-/Å^2$)). Their morphological analysis is often coupled with statistical investigations to produce a 3D reconstruction of their structures. Given their high sensitivity to irradiation by electrons, the focal series necessary for 3D reconstruction can be made only on different areas of the sample. A large number of observations are required to reconstruct the 3D structure of a protein with good resolution. As an example, the optimal resolution of 0.7 nm of a reconstructed 3D structure is reached when between 5,000 and 100,000 particles are analyzed. Thus, the structure of fine particle molecular materials such as viruses, bacteria, and macromolecules (DNA, proteins, etc.) is reconstructed in the same way. The structural information coming from proteins provided by X-rays will be combined with that from TEM analyses to refine the local structures of proteins in proteic complexes.

Protein or DNA crystals or virus films deposited in 2D are considered crystalline materials. In this case, diffraction contrast can be used for imaging.

10.2.2 Structure of Bulk Frozen Samples

In the case of bulk frozen samples, a thin section obtained by cryo-ultramicrotomy can be observed directly using cryomicroscopy in bright-field and low-dose mode (low irradiation of electrons of maximal dose: 1 $e^-/\text{Å}^2$) with a CCD camera at a low light level. This method is the only one that allows access to the structure of a biological material in its hydrated state.

11 Study of Properties

TEM can be used to investigate the local properties of a material in relation to the ordered or amorphous structure in a bulk sample, fine particle, or a multilayer material. Depending on the case, this requires a special specimen holder and/or specific support grid in order to increase the temperature, to measure an electrical current (if a stress must be applied), or if a particular atmosphere is needed.

11.1 Optical Properties

The local optical properties of a compound can be investigated using plasmons in EELS to determine the dielectric constant of a compound. Thus, the local variations of dielectric properties in an oxide can be highlighted near crystal defects such as domain walls. This does not require a particular specimen holder.

11.2 Electrical Properties

The local electrical properties can be measured directly on grain boundaries in conductive ceramics or bicrystal materials. This type of local measurement requires a special specimen holder equipped with four current measurement points on both sides of a grain boundary.

11.3 Electronic Properties

Local EELS analyses, EXELFS, and ELNES in particular are used to determine the electronic structure of an atom in its environment. Thus, the proximity of an interface or crystal defect disturbs the electronic structure (i.e., the chemical bonds) over varying distances depending on the defect and material type. The core of a dislocation can be shown to create a stress around it on a few atomic planes, causing a local change in the absorption threshold structure of the atom at the defect. This type of analysis does not require a particular specimen holder.

11.4 Magnetic Properties

Magnetic properties can be analyzed on a microscope without a magnetic field due to the polar piece of the objective lens. This type of observation constitutes Lorentz microscopy; it makes it possible to highlight the magnetic domain walls with little enlargement compared to classic TEM microscopy. Local magnetic properties can also be studied using electron holography in a TEM microscope equipped with a FEG.

11.5 Mechanical Properties

Mechanical properties can be studied locally in a volume or close to a crystal defect, or from the quantitative processing of parameter variations. In the first case, observations are made using special specimen holders where the sample can be subjected to particular compressive or tensile stresses. In the second case, they involve the quantitative analysis of either HRTEM images or LACBED diffraction patterns to trace parameter variations resulting from relaxation near an interface or in a volume.

11.6 Chemical Properties

Simple electron irradiation under vacuum is used to decrease the crystal–amorphous phase transformation temperature of a simple system such as silicon. Observations under an electron beam are used to investigate the stability of phases, their crystallization, amorphization, transformation with temperature, etc. They do not require a particular specimen holder when working at room temperature.

11.7 Functional Properties

The location of functional sites can be investigated using CTEM microscopy on a biological material labeled beforehand using functionalized antibodies that recognize the antigens present that are being investigated in the preparation. This preparation type does not require a particular specimen holder.

Table 4.4 Sample thicknesses required for structural and physical characterization, depending on the TEM technique

Observation mode	CTEM				TEM/STEM and STEM					
Analysis type	BF/DF (nm)	WB (nm)	SAED (nm)	HRTEM (nm)	Microdiff. Nanodiff. (nm)	CBED (nm)	LACBED (nm)	HAADF (nm)	EDS (nm)	EELS (nm)
Topography	<50									
Structure				<50	<50	≥ 50	≥ 50	<50		
Structural defects	<50	≥ 50	≤ 50	<50	<50	≥ 50	≥ 50	<50		
Crystallography			≤ 50		≤ 50	≥ 50	≥ 50			
Chemical composition								≤ 50	≥ 50	<50
Chemical bonds										<50
Properties	≥ 50			<50		≥ 50	≥ 100			<50

Table 4.5 Different types of analyses correlated with possible observation modes

Observation mode	CTEM					TEM/STEM				
Type of analysis	Imaging BF/DF	Imaging WB	Diffraction SAED	Imaging HRTEM	Imaging Lorentz	Diffraction CBED LACBED Microdiffraction Nanodiffraction	Imaging HAADF	Analysis and imaging EDS	Analysis and imaging PEELS/EELS	Analysis and imaging Holography
Topography	+									
Structure	+	+	+	+	+	+	+	+	+	+
Crystal defects	+	+	+	+	+	+		+	+	+
Crystallography			+			+				
Chemical composition								+	+	
Chemical bonds									+	
Properties	+			+	+	+			+	+

12 Relationship Between Sample Thickness and Analysis Type in TEM and TEM/STEM

One of the characterization limits is sample thickness. Whereas thickness can reach values of 200 nm for bright-field and dark-field images, in the case of high resolution, the maximum thickness is 50 nm for energy-loss analyses. As the optimum is between 5 and 20 nm, high-voltage microscopes are used for such analyses.

However, this upper limit is not as favorable for CBED and EDS analyses which, respectively, require a minimum sample volume to ensure 3D interactions resulting in the formation of HOLZ lines, as well as to have a detectable X-ray output. Knowing the thickness of the sample determines the quantitative chemical analysis of the interfaces. This can be determined using CBED diffraction or using plasmons. Table 4.4 shows the ranges of sample thickness required for the various analysis types.

For materials composed of light elements, the optimal thickness of the TEM sample for observations in bright-field mode is between 50 and 100 nm. Overly thin sections lack contrast, while overly thick sections lack definition. This is because several different structures are mixed in the observation plane. For these materials, it is also preferable to work with beam acceleration voltages between 75 and 120 keV in order to achieve an image contrast as high as possible.

13 Assessment of TEM Analyses

The use of various observation modes (Table 4.5) in parallel on the same area to be analyzed is very helpful. Analyses can be coupled in image and diffraction modes and chemical analysis can be performed for the same spatial resolution (sizes of areas analyzed at 20 Å on an analytical 200 keV microscope). Using the image mode, local structural defects, grain boundaries, intergranular phases, interfaces, segregations, precipitates, etc., can be characterized. Using the diffraction modes, the local structure can be determined at the same time that chemical analysis by EDS and/or EELS is performed. However, it is necessary to take into account that the optimal EDS analysis conditions are different from those of EELS, in particular with regard to the sample thickness. Note that the recommended thickness for EELS analysis is excellent for high resolution; therefore EELS and HTREM analysis can be combined for the same preparation.

The best compromise must always be found for the optimal use of all these techniques, in order to respond to a problem without damaging the material.

Chapter 5
Physical and Chemical Mechanisms of Preparation Techniques

1 Introduction

Microstructural investigations of materials using transmission electron microscopy involve two constraints due to the illumination source. Electrons displace only in a high vacuum, and even when highly accelerated, they transit only a very small material thickness. The sample preparation should resolve both of these issues: the sample must be stable under vacuum and it must be very thin (on the order of 100-nm thick).

Most types of materials are solids and stable under vacuum; this is not true for hydrated samples or those in liquid solution. Water must either be immobilized by transforming it into ice or eliminated. This latter step requires a complex procedure in order to prevent changing the original structure.

The classical preparation techniques for materials studied in TEM thin the sample in the form of either thin sections or a powder made up of very finely dispersed particles whose thickness in the observable area does not exceed 100 nm.

Several preparation techniques involve different interaction mechanisms for obtaining a thin slice that is observable in the transmission electron microscope (TEM). Some of these mechanisms involve the abrasion or dissolution of the material, while others only break the material or immobilize it in a state close to its initial state. In some cases, it is necessary to change the material's properties or physical state; in other cases, physical, chemical, mechanical, or ionic actions are involved. It is often necessary to combine several types of actions during the preparation steps in order to obtain a thin slice.

The mechanical actions involved are rupture or abrasion. An ionic action is an abrasion of the surface of the material by ionized particles. There are two types of chemical actions: dissolution or bridging reactions. One mechanism for immobilizing the structure consists of modifying its physical state through freezing. Lastly, a final mechanism, which can be either physical or chemical, consists of adding particles to enable topographical observation or contrast enhancement. Sometimes, it is only possible to observe a sample replica. Depending on the type of material, its physical state, and its organization, these actions can change the structure and go so far as to transform it. This must be taken into account during TEM analysis.

J. Ayache et al., *Sample Preparation Handbook for Transmission Electron Microscopy*,
DOI 10.1007/978-0-387-98182-6_5,

2 Mechanical Action

2.1 Principles of a Material's Mechanical Behavior

The stresses a material undergoes can result in different responses depending on the material type and the stress, temperature, loading rate, and environmental conditions (water, liquid metal, hydrogen, etc.). One of these responses is deformation and the other is rupture (or fracture). Some ductile materials can change the behavior when the temperature is lowered, becoming brittle.

When a material is stressed, it will deform elastically. During this deformation, the material stores up internal energy corresponding to an elastic energy. When the stress is released, the deformation reverses. When the elasticity limit of the material is reached, it deforms under higher stress and exhibits plastic behavior. This deformation mechanism corresponds to a local restructuring that is often irreversible. These different stages are presented in Fig. 5.1.

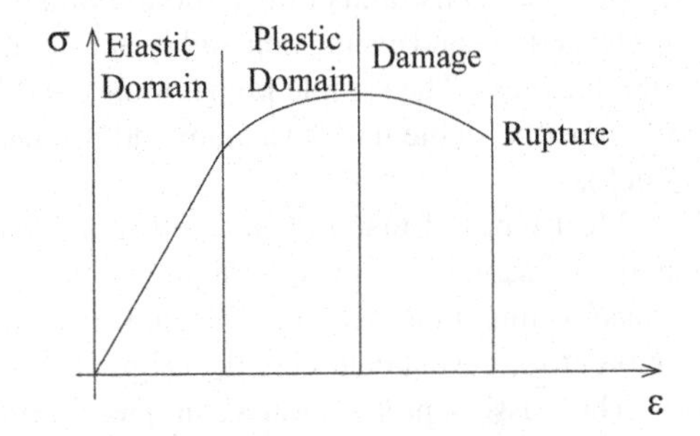

Fig. 5.1 Graph showing the different steps of deformation (*y-axis*) as a function of the stress (*x-axis*)

The plasticity zone contains two domains: a plastic deformation area and a damage area. There are two methods of passing from elastic deformation to plastic deformation: Either there is a smooth transition and no clear distinction can be seen between the linear part and the plastic part (ductile samples) or there is a recess at the start of the plastic deformation, the stress falls, and an irreversible defect is created (brittle samples). This is called the damage area. Cracks in a material appear at the end of the plastic deformation curve. In brittle materials, the damage rapidly causes a rupture, whereas in ductile materials, the plasticity area can vary significantly up to superplasticity.

In brittle materials, there is no macroscopic plastic deformation. The propagation of cracks is very rapid and the rupture is neat, without a reduction of the localized cross section (contraction) following the crystallographic planes. Plastic deformation quickly causes damage, creating defects. This is the case for ceramics, minerals, a large number of vitreous thermoplastics above their glass transition temperatures (T_g), thermoset polymers, and elastomers with low cross-linking rates. In the field of biology, materials such as bone, tooth, and siliceous or calcareous skeletons present this type of brittle structure. Brittle fractures can be transgranular, intergranular, or interatomic.

In ductile materials, the material deformation is first linear (the elastic domain), then it bends (plastic domain) before decreasing (domain of damage and rupture). In an amorphous or disordered environment, there is no lattice, and therefore no dislocations. However, there are plastic deformations. For this type of material, the irreversible deformations may be considered to correspond to a local restructuring. When the stresses exerted on one of these areas exceed a threshold value, the atoms or molecules rearrange themselves locally, resulting in a redistribution of the elastic stresses in the material as a whole. Other zones in the material will in turn rearrange themselves, causing a domino effect. Plastic deformation occurs at a constant volume, a bit like if one slides playing cards over one another (Fig. 5.2).

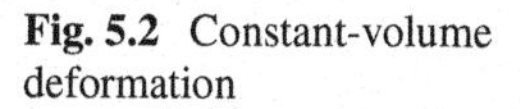
Fig. 5.2 Constant-volume deformation

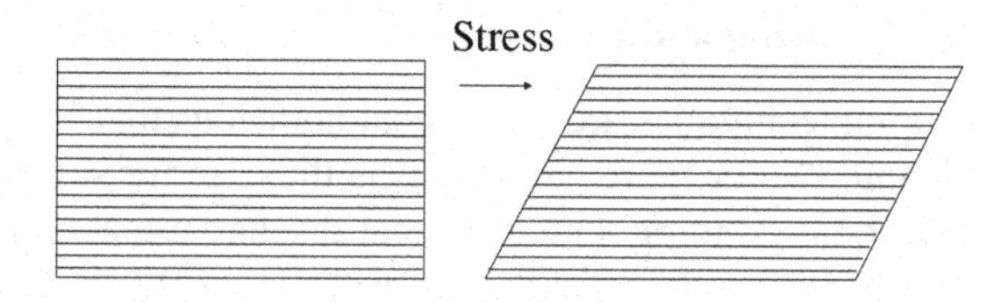

Plastic deformation is irreversible. Residual deformation still remains even when the stress is relaxed.

Ductility refers to a material's ability to deform itself plastically without breaking. Rupture occurs when a defect (crack or cavity) induced by the plastic deformation propagates itself. Therefore, ductility is a material's ability to resist this propagation. Soft materials, some amorphous thermoplastics, as well as most non-crystallized biological materials fracture in a ductile manner.

2.2 Abrasion Principle

Mechanical abrasion corresponds to rubbing a tool against a sample to cut it, to reduce its thickness, or even to polish it. Unlike the fracture mechanism, which results in sample rupture, the abrasion mechanism is still within the domain of plastic deformation (Fig. 5.1). Nevertheless, it creates defects in the sample. This damage results from the application of a stress, corresponding to irreversible plastic damage (strain hardening, dislocations, compression, and creation of cracks).

Abrasion always involves the loss of matter from the sample. The mechanical action of abrasion corresponds to a loss of matter from microcracks or fissures in a material and is caused by a tool. This type of preparation involves three components: the material, the abrasive, and the lubricant. The determinant parameters of this physical action are the stress applied, the rate of erosion, and the nature of the tool.

The abrasion mechanism causes damage at different depths in the material. Depending on the properties of mechanical material, the damage caused to it may reach a depth equal to three times the grain size of the abrasive used.

The damage caused is less significant in brittle materials than in ductile materials. This damage also has a varying degree of intensity depending on the stress applied, the type and grading of the abrasive used, the rubbing speed, and the temperature induced in the material during the process. The stresses initiated by abrasion can cause different types of defects to form in the material: dislocations, strain hardening, twinning, microfractures, fissures, and the dispersion of gaps, cavities, etc.

There may also be thermal effects that result in the localized re-fusion of the material, its transformation, and a phase migration or loss. Lastly, chemical changes due to pollution, matter transport, or chemical-element diffusion may occur.

2.2.1 Techniques Involving Cutting by Means of Mechanical Abrasion: Sawing and Grinding

Cutting techniques are directional abrasions that lead to the separation of the sample into two or more parts. This is the case for sawing, cutting with a cutting tool, ultrasonic grinding, and electrical-discharge machining (EDM). The stresses initiated by abrasion cause cracks at different depths in the material, up to complete cutting.

Sawing can be performed with a disk or wire, of which either the whole or outside (internal or external) contains abrasive grains of different types (diamond, carbide, etc.) and varying gradings. The use of a lubricant (water, oil, etc.) helps to eliminate abraded grains and limit temperature rise during the process.

Ultrasonic cutting is performed using a hollow cutting tool in the necessary shape. This tool is attached to an ultrasonic generator. Abrasive grains in solution are placed on the sample surface, and the tool is moved into contact with the material. The tool is vibrated (and therefore the abrasive grains are vibrated as well), and the application of pressure helps to cut the material.

The experimental conditions chosen depend on the characteristics of the tool used (wire saw, wheel saw, cutting tool with or without ultrasound, etc.) and the nature of the sample.

Cutting is a preliminary preparation technique inducing defects that must be eliminated by the thin slice preparation technique. However, the more defects introduced in the sample at this level, the more difficult it is to eliminate them during final preparation.

2.2.2 Abrasive Techniques: Mechanical Polishing, Dimpling, and Tripod Polishing

The process of abrasion uses abrasive grains of decreasing grain sizes down to below 0.025 μm, in order to minimize material surface roughness.

Abrasion by mechanical polishing generally uses polishing disks containing carbide grains (SiC, ZrC, etc.) of decreasing size (e.g., 60, 30, 12, 7 μm) that are embedded in the support. Carbide abrasion is followed by diamond abrasion, either using diamond pastes or diamond sprays, also of decreasing grain sizes (6, 3, 1, 0.25, 0.1 μm).

Dimpling uses the same diamond pastes as those used for mechanical polishing. The final polishing can be done using colloidal silica or another colloid of a very small grain size.

The tripod technique uses supports in which diamond (alumina, etc.) grains are embedded within the support and are therefore immobile. This enables better quality unidirectional polishing.

Even when the best polishing conditions are chosen, such as those used in the tripod technique, there are always residual microcracks at the last level of mechanical polishing. These are to be eliminated, if the type of material so allows, through the use of an additional ionic or chemical technique to produce a thin slice lacking defects visible in the TEM.

2.3 Rupture Principles

Unlike abrasion, rupture does not involve any loss of sample matter. It involves a fracture induced within the sample, resulting in the sample separating into two or more parts. In the case we are interested in, a tool induces the crack.

Rupture is heavily influenced by the presence of internal defects such as microcracks, pores, inclusions of fragile particles, material heterogeneity, and notches (microcracks resulting from manufacturing or design defects).

Fracture involves three phases: a crack-initiation phase, a crack-propagation phase, and a sample-rupture phase under the effect of stored elastic energy.

Through geometric focusing, the crack causes a high concentration of stress at its end that is capable of breaking the atomic bonds. Rupture mechanics show that this phenomenon depends on two magnitudes: the stress applied and the dimensions of the microcracks. These two magnitudes are combined in the stress intensity factor. For a single material, the critical stress can vary depending on the dimensions of the microcracks present in the material. The rupture occurs when the critical value is reached.

Depending on the preparation technique, the crack initiation is what will vary; the propagation of the crack and the rupture develop naturally from the application of a stress.

In the case of crushing, the start of the crack is produced by the crushing of the material between a mortar and pestle. In the case of cleavage and sectioning by ultramicrotomy, the start of the crack is produced by introducing a tip (or knife blade) into the edge of the sample.

In the case of a freeze fracture performed with a double cupel, the start of the fracture is caused by opening the cupel.

2.3.1 Techniques Involving Fracture: Crushing, Wedge Cleavage, Ultramicrotomy, and Freeze Fracture

Crushing applies to brittle materials or materials made brittle by cooling. It involves applying a stress powerful enough to cause the material to rupture *in random directions* and into a very large number of fragments. In the case of crushing to obtain a thin slice for the TEM, the residual grains must be small enough to present transparent fragments to the electrons, either over all of the grains or on their extremities. Depending on the material crushed, the fracture can be transgranular, intergranular, or interatomic. Liquid nitrogen cryogenics is used to harden a ductile or soft material to make it brittle and capable of being crushed.

Wedge cleavage is the result of an interatomic separation along crystallographic planes with low Miller indices. A single crystal can cleave according to different crystal planes of high atomic density. The start of the fracture generates a high dislocation rate, then a fracture occurs and propagates throughout the crystal along an atomic plane. Figure 5.3 shows the cleavage planes of gallium arsenide (GaAs).

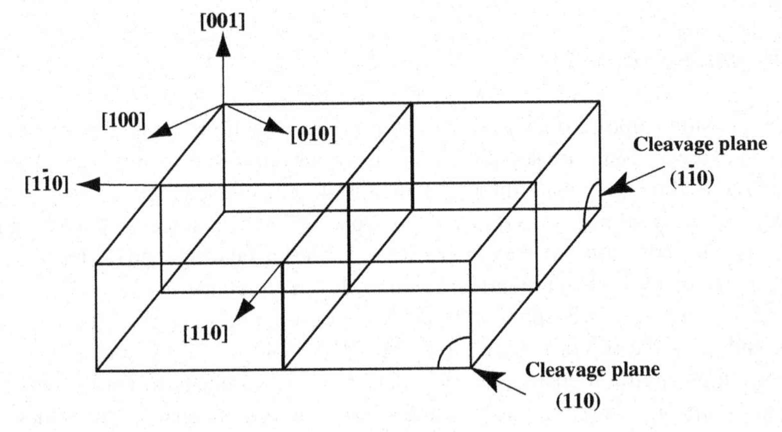

Fig. 5.3 Cleavage planes (110) and (1–10) of a GaAs single crystal relative to the crystallographic directions

Ultramicrotomy and cryo-ultramicrotomy use a knife as a cutting tool. The crack is started by the cutting edge of the knife, which is in the shape of a prism. The increasing thickness of the knife opens the crack and acts as a stress on this crack. The crack propagates within the sample because of areas of low cohesion and material heterogeneities. The most fragile bond in the material breaks first. The fracture is directional, corresponding to the introduction of the knife (Fig. 5.4).

Distortion forces are involved during cutting. In Fig. 5.5 we see that cutting force R under knife angle θ has divided into two components, force F_c, which causes the fracture, and force F_t, which causes compression of the material. This compression is greater if the knife angle is large. This angle can vary between 35° and 55°. The section fractures into small fragments if the material is brittle and is compressed if the material is soft.

For cryo-ultramicrotomy, the mechanical action is the same type as that described for ultramicrotomy. Since the material is frozen, this is an ice-based heterogeneous compound, which behaves like a brittle material.

The freeze-fracture technique is applicable to multiphase samples that have an aqueous phase. Water is transformed by freezing into the solid phase. Frozen water

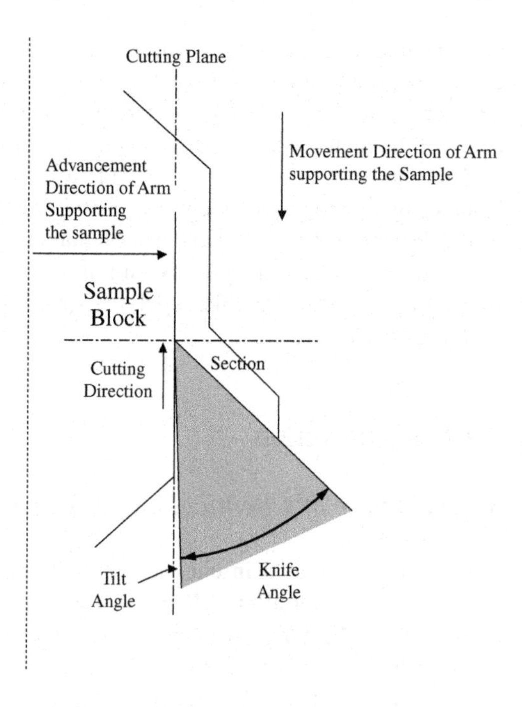

Fig. 5.4 Geometry of the knife and the sample block

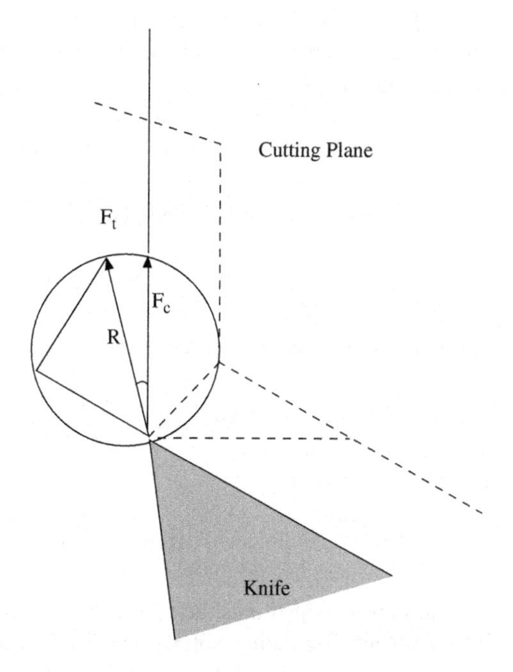

Fig. 5.5 Forces involved during cutting

behaves like a brittle sample, since vitreous or crystalline ice has little plasticity. Because of the heterophase nature of the material, rupture will occur in the areas of least cohesion. For biological materials, it will occur at the level of the non-polar zones of the membranes, causing an opening of the double phospholipid layer.

The start of the fracture can be made in two different ways, either using a razor blade or by opening a double cupel containing the sample. In the case of the razor blade, the start is directional and a particular orientation can be selected. The rupture is similar to that described in cryo-ultramicrotomy. In the case of the opening of a freeze fracture with a double cupel, the start is random within the material and the rupture is due to a tear.

3 Chemical Action

3.1 Principle of Chemical and Electrochemical Dissolution

The process generated in chemical dissolution is a reduction–oxidation reaction (redox) of the material in a chemical solution. An electrical voltage spontaneously establishes itself at the material–solution interface. This is enough to induce material dissolution. It is a surface reaction that brings ions and electrons into play, and therefore comes close to an electrochemical polishing. Dissolution rates can be very high, from 50 to 500 μm/min.

Electropolishing results from an anodic dissolution (reduction–oxidation reaction) of the sample surface under controlled potential. The processes of anodic dissolution are characterized by their complexity and by the wide variety of factors affecting the process.

The following reaction may occur, depending on the nature of the metal dissolved, composition of the electrolyte, temperature, and current density:

(1) Transfer of metallic ions and electrons in the solution

$$\mathrm{Me} \rightarrow \mathrm{Me}^{2+} + 2\mathrm{e}^-$$

(2) Formation of an oxygen layer

$$\mathrm{Me}^{2+} + 2OH^- \rightarrow \mathrm{MeO} + H_2O + 2\mathrm{e}^-$$

(3) Oxygen release

$$4OH^- \rightarrow \mathrm{O}_2 + 2H_2O + 4\mathrm{e}^-$$

The sample constitutes the anode and a conductive plate constitutes the cathode. The generally oxidizing solution is used to dissolve the surface of the conductive material. At the solution–metal interface, a reactive surface layer is formed, regulating the dissolution of the material. The electrolytic solutions used generally contain three components: an acid or a base, in order to ensure oxidation of the material; an agent that regulates the speed of dissolution (passivator); and a final element that promotes a viscous state in the solution–metal interfacial zone. This viscous layer becomes loaded with dissolution products and the result is a layer whose electrochemical properties, as well as its viscosity, are modified. A concentration gradient forms from the crevices to their projections, i.e., in the different areas of roughness on the material's surface. There is then a higher current density on the projections, which dissolve faster than the crevices because the current flow is facilitated and therefore the current density is higher. The surface gradually levels out. This preferential dissolution eliminates the sample roughness caused by mechanical polishing.

Several theories (Jacquet, Elmore, Edwards, Levin, Batashev and Nikitin, Shigolev, etc.) explain this effect, but none of them are regarded as a universal theory. Figures 5.6 and 5.7 show a model of the mechanism involved in this electrolytic abrasion.

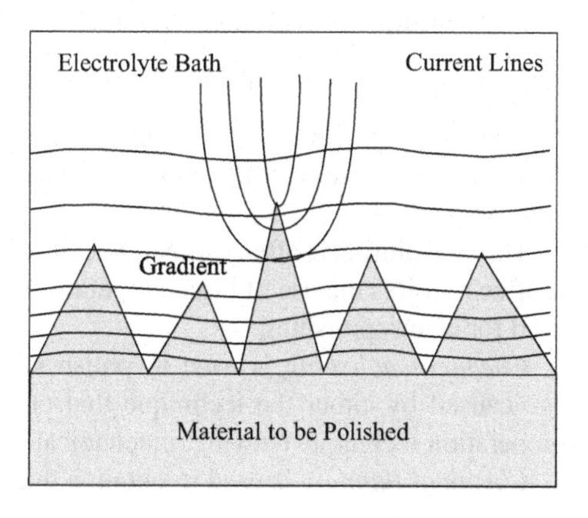

Fig. 5.6 Current density is proportional to the concentration gradient; it is lower in the crevices and higher in the projections

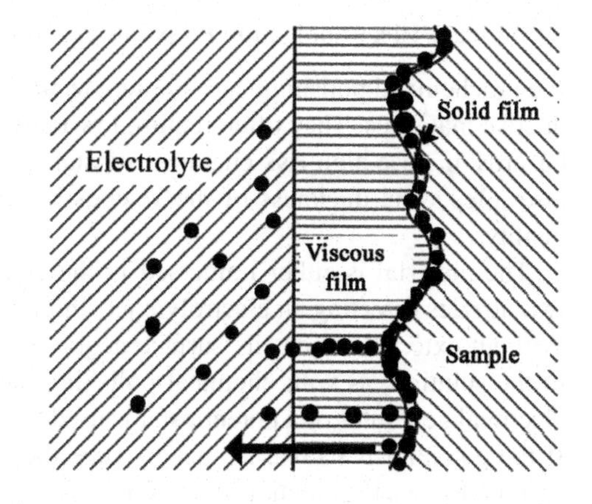

Fig. 5.7 Electrolytic polishing principle (according to Jacquet). The projections dissolve faster than the crevices

The selection of commercial polishing solutions is fairly limited; it is often necessary to adjust the composition and concentration of components depending on the metal or alloy. The electrochemical conditions (potential and current) must be customized for each polishing cell. Indeed, they depend on the geometry of the cell and the respective positions and the size of the cathode and anode electrodes (sample). The cathode must always have a large surface area compared to the anode and must be positioned across from and near the anode.

The tables available in specialized publications provide a large number of chemical compositions and working conditions (type of cathode, temperature, polishing potential, and current density) that can be adjusted.

There are three classes of electrolytes for processing most metals:

1. Perchloric acid (1–20%)/ethanol
2. Perchloric acid (5–20%)/other acids
3. Chromic acid/acetic acid solutions

3.1.1 Techniques Involving Chemical and Electrochemical Dissolution

Preliminary Chemical Polishing Preparation and Chemical Thinning Technique

The chemical polishing and thinning techniques are used to thin non-electrically conductive materials (semiconductors, glasses, oxides, etc.) or conductive materials that can be dissolved chemically.

This dissolution is often difficult to control, in particular for multiphase materials. It often requires the use of high temperatures and more reactive solutions than those used for electropolishing.

Chemical polishing is used to polish surfaces in order to eliminate any damage caused by either the technique that produced the material or the preliminary preparation technique (sawing, mechanical polishing, etc.).

Chemical thinning is used to obtain a thin slice.

Preliminary Electropolishing Preparation and Electrochemical Thinning Technique

Electrochemical polishing and thinning techniques are used to thin electrically conductive materials. They often require low temperatures to prevent explosions or fire hazards, in case high-risk chemical substances are used. Lowering the temperature also affects the thickness of the viscous layer, controlling the polishing rate and quality.

The material is either immersed in the electrolytic solution or placed between two nozzles directing an electrolyte jet simultaneously at both sides of the material. Without external action, its surface acquires an equilibrium potential with regard to the solution, as with chemical polishing. Under electrochemical conditions, an electrical generator or a potentiostat is used to set and regulate the surface potential. If the potential is increased successively, the surface acquires a stable dissolution current for each potential value. Thus, a curve called the current/voltage curve can be plotted. Recordings can be used to observe an increase in current using low potentials (cathodes with hydrogen release), which reaches a plateau (called the "polishing plateau"). Then, the current increases again and there is an oxygen release. Before the plateau, the surface presents localized etching on the high density crystal planes; after the plateau, small holes can be observed. A good polishing condition is situated at the last third of the plateau, where the apparent resistance of the electrolytic cell is greatest.

What is happening? The anions of the solution lose their hydration water around the anode (metal) and are adsorbed at the electrolyte–metal interface, forming a viscous layer. Often, the anions combine with the metal to form crystal salts that can be observed in the microscope using polarized light. They disappear once the current is interrupted.

Electropolishing is used to polish surfaces, thereby eliminating any damage caused by the material production technique or the preliminary preparation technique (sawing, mechanical polishing, etc.).

Electrochemical thinning is used to obtain a thin slice.

4 Ionic Action

4.1 Ionic Abrasion Principles

Ionic abrasion results from the interaction between the material and ionic particles. Ions created by an electrical discharge are accelerated under a few kiloelectron volts (0.5–6) in a focused beam with a Gaussian current density. This ion beam is directed at the surface of the sample, in the area to be thinned. When an ion meets a material surface, it penetrates the material until it successively hits different atoms (or ions in the case of ionic compounds) with sufficient energy to displace them. In turn, the atoms of the lattice will be projected into the solid, resulting in new collisions that cause the atoms on the surface to tear; this is the phenomenon of pulverization. The output of pulverization is characterized by the ratio of the number of atoms torn to the number of incident ions.

Under normal incidence, the pulverization output depends on the energy of the incident ion. Below a certain energy level (10–40 eV) no atoms are expelled. The output increases to reach a maximum of between 2 and 30 keV, a range in which an ion tears between 1 and 50 atoms (this depends on the nature of the target and the energy of the ions). If the energy is too strong, the ions enter the material without tearing atoms. There is an implantation of ions and this occurs more if the material is composed of light elements. In an initial approximation, the pulverization output, below the maximum, is proportionate to the logarithm of the acceleration voltage (Fig. 5.8).

The higher the mass of the incident ion (rare gas or metal), the stronger the pulverization output. The higher the index of the crystal plane, the faster the tearing speed, but the maximum on the curve is displaced toward high energies because

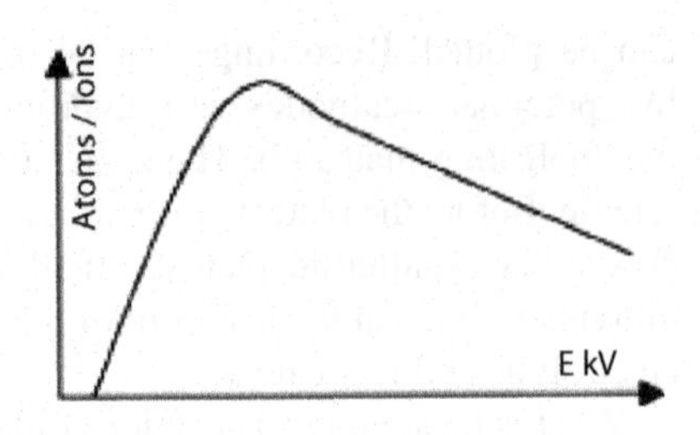

Fig. 5.8 Yield of pulverization as a function of acceleration voltage

there is greater resistance. The flow of incident ions decreases with the cosine of the angle incidence.

The penetration depth of the ions in the material is proportional to the acceleration energy of normal incidence and inversely proportionate to their mass. With argon ions, it is 10 nm under 6 keV with an incidence angle of 70°, but falls to 1 nm in low-angle incidence.

The process generates many artifacts as follows:

- The implantation of ions in the depth of the thin slice, the creation of a thick layer (some tens of nanometers) of material rendered amorphous (destruction of the crystalline network) on sample surfaces.
- A strong sample temperature increase (depending on ion milling parameters and whether or not the sample can be cooled), which may induce a change in the stoichiometry, phase transformation, demixing, or decomposition of the material.
- Possible differential scouring, causing differential component sputter rates (roughness), resulting from the difference in hardness and atomic weight between these components, differences in crystallographic orientations of the material, and the incidence angle of the ion beam (Fig. 5.9).

The use of a low voltage, low current, and minimal incidence angle can reduce damage; however, this drastically increases the thin slice preparation time.

This type of abrasion is used to polish any type of hard and soft, single-phase or multiphase materials.

4.2 Techniques Involving Ion Abrasion

4.2.1 Ion Beam Thinning and Focused Ion Beam Thinning (FIB)

Ion beam thinning is performed using equipment under a vacuum on the order of 10^{-4}–10^{-5} Pa.

Classical ion guns are composed of an ionization chamber and acceleration electrodes (0–15 keV). Focusing the beam enables a precise area of the sample to be thinned, on the order of a millimeter. Often, sample cratering is caused until a central hole is formed.

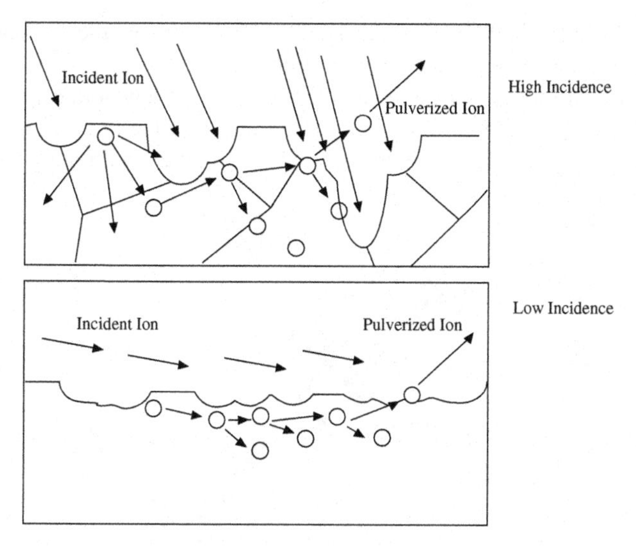

Fig. 5.9 Selective etching as a function of the nature of the material and the angle of incidence of the ions

The gas used is generally an inert gas, preventing any chemical reaction with the material. The gas must also have a high atomic weight to be able to pulverize heavy atoms. Argon (Ar), with an atomic mass of 39.9 amu, is frequently used. It is also an element that is rarely found in the chemical composition of materials, which helps to easily identify its implantation in the sample.

In some cases, it is preferable to use a mixture of an inert gas and a reactive gas in order to accelerate abrasion or reduce artifacts generated by the usual argon technique. For example, an argon–oxygen mixture can be used to thin diamond, and iodine can be added to eliminate drops of indium that form on the surface of a material containing indium during argon thinning.

The ion acceleration voltage is between 100 eV and 10 keV. The use of high voltage (and strong current) helps to rapidly abrade the material (up to 100 μm/h), but generates a large number of artifacts. Applying voltage as low as possible (depending on the type of available equipment) can markedly limit these artifacts. During treatment, it will be applied only at the end of thinning because of the very low rate of abrasion resulting from it. The use of a liquid-nitrogen-cooled specimen holder reduces the temperature rise in the sample.

Two ion guns are generally used. These guns can be situated above, below, or on both sides of the sample. The incidence angle between the ion beam or beams and the sample plane can be from 0° to 90°, depending on the equipment. The angles generally used are between 1° and 15°. As with acceleration voltage, using the lowest angle possible will considerably reduce artifacts, but will increase the time needed to obtain a thin slice.

Some instruments can be used to apply a current (delaying potential) within the sample via the specimen holder. This arrangement can divert the incident electrons at the surface of the sample by a tangential arc effect, reducing their implantation in the material and reducing surface layer amorphization.

The ideal conditions would be the continuous use of a low voltage, an incidence angle near 0°, and sample cooling, but the ion milling time in this case would be extremely long. Figure 5.10 illustrates the principle of the ion thinning technique.

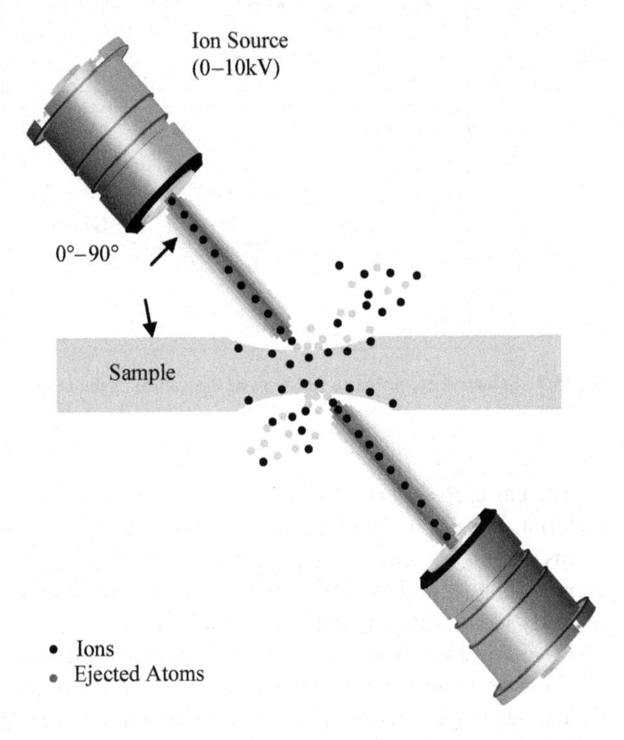

Fig. 5.10 Diagram of the ion thinning principle

Focused ion beam thinning (FIB) is generally performed in a scanning electron microscope (SEM), coupled with an ion gun (Fig. 5.11) equipped with electrostatic lenses, under a vacuum of approximately 10^{-6} Pa.

The ion source used for the FIB thinning technique is generally gallium (Ga), whose atomic mass is 69.7 amu; however, other metals can also be used: Au, Be, Si, Pd, etc. Gallium has multiple advantages:

- Its very low fusion point (302.8 K) helps to minimize the interdiffusion reaction between the liquid and the tungsten needle, along which the gallium "flows."
- In the case of chemical analysis (EDS), the energy peaks of gallium do not overlap much with the energy peaks of other materials.
- It has excellent mechanical and thermal properties.

The ion beam is generated from a gallium reservoir in contact with a very fine tungsten (W) tip. Liquid gallium "flows" along the tungsten tip, and an electrical field (10^8 V/cm) applied to the tip of the tungsten needle causes the emission of

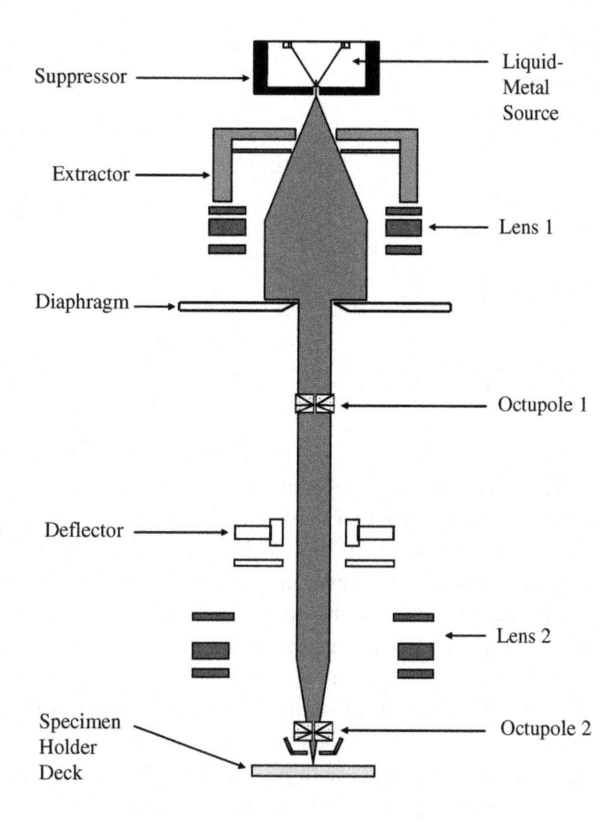

Fig. 5.11 Diagram of a type of FIB column

positive ions, which are extracted and accelerated by a voltage of a few kiloelectron volts (Fig. 5.12). The current density is adjusted by selecting the current of the electrostatic lenses and the size of the aperture. These last two parameters determine the diameter of the ion beam, which can be less than 10 nm. Material abrasion occurs when the ion beam is scanned over the area of interest (Fig. 5.13).

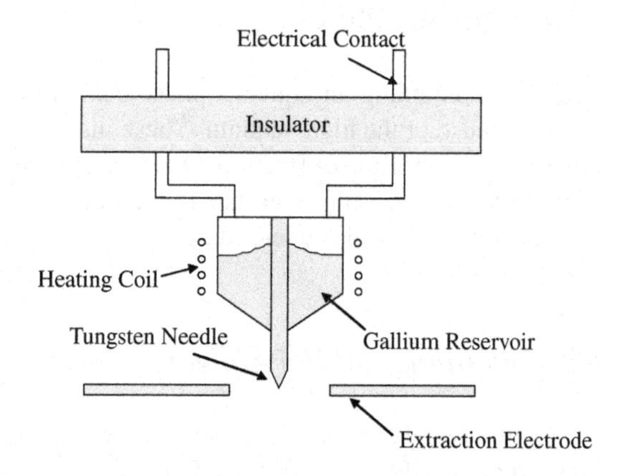

Fig. 5.12 Diagram of the gallium source

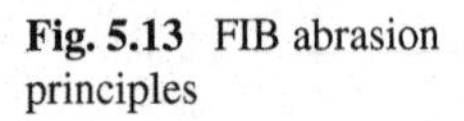

Fig. 5.13 FIB abrasion principles

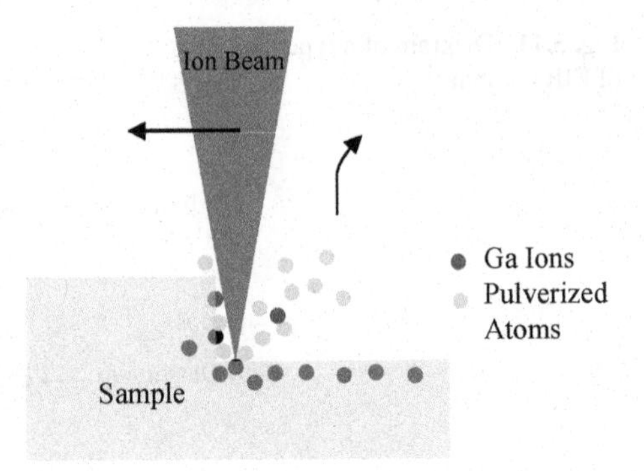

The acceleration voltage of the ions is between 1 and 30 keV, and the current density is on the order of between a few tens of picoamps and several hundreds of nanoamps. The use of a high voltage and a strong current is required to quickly abrade the sample. However, as with the ion beam thinning technique, this induces significant artifacts that are substantially the same. These artifacts must therefore be reduced through the use of low voltage and a weak current.

The incidence angle between the beam and the sample is adjusted by tilting the sample. It is generally 0° with regard to the plane of the future thin slice for quick abrasion of the material and can be varied between 0.5° and 2° for the final thinning step and for "cleaning" the thin slice. The introduction of an angle in the final step is necessary to compensate for the non-homogenous thickness caused by the Gaussian distribution of intensity in the ion beam.

The use of a beam with a very small diameter and an angle of incidence near zero limits preparation artifacts.

5 Actions Resulting in a State Change of Materials Containing an Aqueous Phase

Materials containing an aqueous phase cannot be observed directly in the microscope because of the high vacuum. These materials are also usually soft and cannot be thinned by abrasion or fracture. One or more prior preparations must be applied in order to modify their physical state for preparation and observation. This is generally the case for biological organic samples.

5.1 Elimination of the Aqueous Phase

There are several methods for removing the water contained in a material. The water phase diagram (Fig. 5.14) shows that there are two ways to change to the gaseous

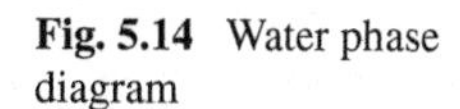

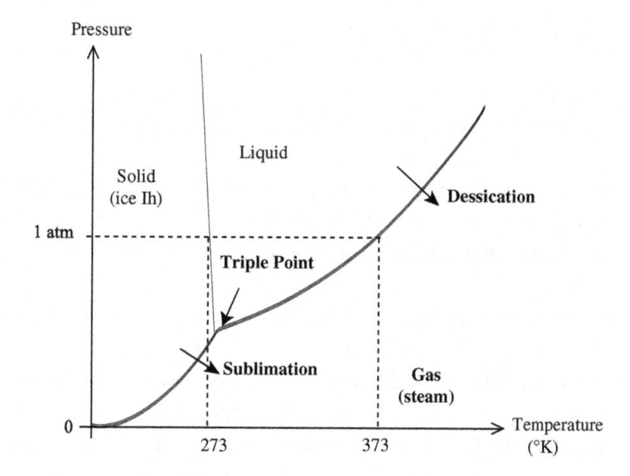

Fig. 5.14 Water phase diagram

phase, either by desiccation (passing from the liquid phase to the gaseous phase) or by sublimation (passing from the solid phase to the gaseous phase).

Desiccation is the simplest method; it consists of letting the water evaporate in the open air at room temperature, but this method destroys delicate structures because it is far from equilibrium between the vapor tension of the water and the pressure of the water vapor that dissipates into the atmosphere. Desiccation can be conducted correctly only at the critical point (where these two pressures are equivalent). For water, the critical point is very high in temperature (647 K) and pressure (221×10^5 Pa), and therefore is technically difficult to reach.

Sublimation consists of making ice evaporate. It first assumes the transformation of water into ice through freezing, and then sublimation of the ice. Since the surface tension of ice is very poor at low temperatures, it is necessary to apply a strong depression.

There is a third option that consists of replacing the water with a solvent that can be mixed with water: ethyl alcohol, methyl alcohol, or acetone. This technique is called dehydration. The solvent can be allowed to evaporate, because its critical point is easy to reach. Also, the solvent can be replaced (substituted) with a resin that is then hardened.

Therefore there are two possible approaches for treating hydrated samples:

1. Saving the water and transforming it into ice, following a purely physical procedure. The material is then directly cut by cryo-ultramicrotomy, or cryo-embedding is done. To carry out the latter, the ice must be eliminated beforehand by cryo-substitution (freeze substitution) or by cryo-sublimation (freeze drying).
2. The water is eliminated and the sample is embedded in order to make it solid; this is a procedure that employs the contribution of chemical molecules, which will be seen later.

5.2 Freezing Principles

Freezing is the transformation of water into ice, i.e., the passage from the liquid phase to the solid phase. This state change is the result of two phenomena occurring simultaneously: nucleation and diffusion. Nucleation is the production of small ice-formation sites, constituting the seeds of the first crystals. Diffusion is the ability of water or ice to move. The seeds have a significant attraction to the free water or ice molecules that migrate and increase the size of the crystals.

Water's particularity comes from its molecular structure. In a water molecule, an oxygen atom is bound to two hydrogen atoms by covalent bonds involving a pair of electrons. The O–H–O angle is in the neighborhood of 104.74°. The molecule is electrically neutral but is polarized. Indeed, the density of electrons is greater near the oxygen nucleus than near the hydrogen nuclei. When two water molecules are present, they tend to unite through an electrostatic bond between a positively charged hydrogen nucleus and the electron cloud surrounding the oxygen nucleus. This bond, called a hydrogen bond, tends to make the H–O–H alignment. At a given moment, a water molecule is surrounded by four neighboring molecules, which are ordered according to a tetrahedral geometry as a result of the hydrogen bonds. Liquid water is constantly reorienting itself, but has an instantaneous short-range order. If the temperature falls below 273 K, the tetrahedral structure stiffens and becomes a long-range order, constituting a solid with a hexagonal crystal structure. Homogeneous nucleation occurs. Since water is very mobile, the nucleation sites will get larger in order to form large-dimensional crystals (Fig. 5.15). This is ordinary hexagonal ice (I_h).

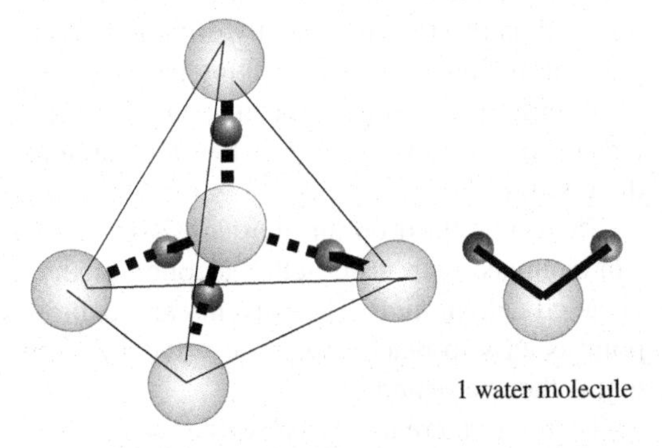

Fig. 5.15 Two possible arrangements of a water molecule

In the solid–liquid system created at the start of the phase transformation, the crystals are unstable and are very quickly redissolved, contributing calories to the system due to their fusion; this is then a metastable state. It is necessary to fight this caloric contribution and surpass this surfusion stage and the eutectic point (corresponds to the temperature at which the largest crystals are formed) by performing quenching at a very low temperature.

Good freezing of water depends on several factors, including two important ones: the temperature at which the state change occurs and the rate of cooling of the system. These factors will determine the size and nature of the seeds formed and reduce the possibility of the migration of the water. This will block an increase in the size of the primordial crystals. A low temperature promotes the appearance of many nucleation sites and maximally reduces the diffusion of not-yet crystallized water before obtaining a change in the state of the whole liquid volume. If in the end the nucleation seeds remain a very small size (not visible at the TEM scale) and do not provide a diffraction pattern, this is referred to as vitreous ice. Vitreous ice is produced at temperatures lower than 77 K. Once the seeds at the nucleation centers start to get larger, the ice becomes cubic crystalline ice.

The vitrification of pure water is obtained through a decrease in temperature, at a rate of at least 10^9 deg/s, to a temperature less than 77 K.

This state change also depends on the pressure applied to the system when the temperature is decreased. A strong increase in pressure (21×10^{-7} Pa bars) promotes the formation of vitreous ice, even with less rapid cooling. The increase in pressure increases the viscosity of the water, and therefore slows the water's diffusion and decreases the possibility of the nucleation seeds growing larger. The vitreous ice obtained is amorphous "high-density" ($d = 1.17$) ice, whereas the ice obtained under normal pressure is amorphous "low-density" ($d = 0.94$) ice. Depending on the temperature decrease, the speed of this decrease, and the pressure exerted on the system, different forms of crystalline ice can exist. There are at least 12 forms.

In the case of organic systems, the material is not in pure water, but rather in solutions and suspensions. Therefore, both free water and water are bound to the molecules. There are fewer tetrahedrons, and there is more space between them. The presence of solutes promotes the appearance of nucleation seeds before the nucleation of the tetrahedrons. Water is immobilized by adsorption onto the molecules, which presents the constitution of a large-scale order and stabilizes the system by minimizing diffusion of the water. In this case, nucleation is heterogeneous. Cubic-type crystalline glass occurs at temperatures near 200 K. In these systems, vitreous ice will be obtained under less extreme conditions. We consider that, for a biological system having an osmolarity of 300 mOsm (e.g., mammals), vitreous ice can be obtained on a very thin film of a solution with a temperature decrease of 10^6 deg/s down to approximately 125 K. Freezing under these conditions is referred to as ultrarapid freezing (Fig. 5.16).

Like water, ice is capable of migrating if the frozen system is not maintained at a temperature low enough to prevent the diffusion phenomenon. For pure water above 140 K, ice crystals are able to become larger, slowly at first, and then quickly as the temperature rises to around 200 K. Then cubic ice crystals appear. This is in the neighborhood of the T_g of vitreous ice (between 120 and 140 K). In a biological-type heterogeneous system (whose diffusion rate is low), experience shows that this phenomenon occurs around 183 K.

Once water is frozen in the best possible way, one of three preparation pathways can be carried out to observe the sample.

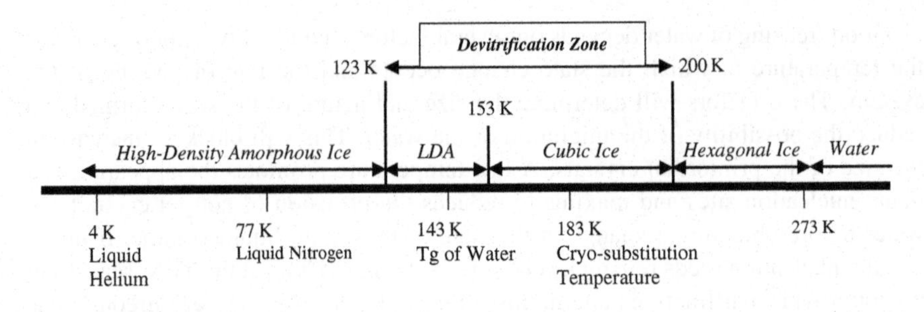

Fig. 5.16 Freezing diagram of an organic solution of 300 mOsm in water

If the sample is made up of small fine particles, it can be observed directly in the vitreous ice layer using a cooled specimen holder on the microscope (in the second volume "Techniques", Chapter 6, Section 2). If it is a bulk sample, it must be sliced. We can obtain thin slices directly by means of cryo-ultramicrotomy ("Techniques", Chapter 4, Section 5). Otherwise the sample must be embedded in a low-temperature polymerizable resin in order to harden it. In this case, the sample must be dehydrated or desiccated beforehand. Dehydration occurs at low temperatures by substituting ice in a solvent; this is called cryo-substitution. Desiccation is carried out by means of sublimation (freeze drying). They will be followed both by infiltration and inclusion, also at low temperatures (see Prior Preparation in "Techniques", Chapter 2, Section 10).

5.3 Principle of Substitution, Infiltration, and Embedding in Cryogenic Mode

In the case of frozen samples, substitution occurs at low temperatures, and ice is gradually replaced with alcohol or acetone. It is important to avoid returning to the liquid phase of water in order to prevent the migration or redistribution of ions, which occurs in this state. It is also necessary to prevent the transformation of vitreous ice into cubic or hexagonal ice because it destroys the structures. Beyond 183 K, this transformation is exponential.

So that substitution can be carried out within the proper timeframe, this operation is performed under vacuum. The procedure is begun at 183 K. The ice is extracted by the vacuum and dissolved in the solvent; the solvent slowly penetrates the core of the sample. This phase must be long enough to extract a good portion of the ice and prevent the possible size increase of the crystals that occurs more quickly once a temperature of approximately 203 K is reached. After a certain period, the temperature is increased regularly in steps so as to ensure total ice substitution. The rapidity of these exchanges is tied to the increase in temperature.

The degree of dehydration required depends on the type of analysis planned. For example, dehydration is stopped at 213 K in a chemical analysis in order to prevent ion migration. For structural analysis or for the immunolabeling technique, dehydration can be increased from 233 to 273 K without risk, depending on the type of antigen to be investigated.

All of the protocol variants can be designed around these values. The resins used for low-temperature embedding are acrylic resins, in complex blends, which are relatively fluid at these temperatures and can polymerize in the cold under ultraviolet radiation.

Lowicryl resins were specially developed for this type of use. Two of them are non-polar (HM20 and HM24) and two are polar (K4M and K11M). They have a wide range of use. Figure 5.17 shows their usual temperatures of use.

The addition of a fixative (OsO_4 or aldehydes) in the cryo-substitution solvent helps to accelerate the process. The presence of a fixative locks the organic molecules and the water bound to them. The temperature can be raised faster without the risk of creating cubic ice. In this case, dehydration can be stopped at 293 K and the sample will be embedded in classic epoxy or acrylic resins.

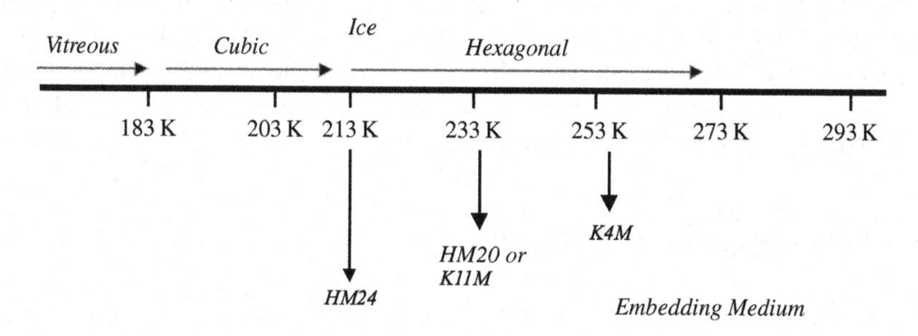

Fig. 5.17 Lowicryl temperatures of use

5.4 Cryo-sublimation (or Freeze-Drying) Principle

The vapor pressure of ice is tied to temperature and pressure. When the temperature is increased, and therefore the vapor pressure is increased, the evaporation of the ice is accelerated. The pressure must also be decreased in order to achieve the same effect.

Since sublimation must occur at a low temperature (183 K) in order to prevent the recrystallization of vitreous ice into cubic ice, sublimation occurs under a vacuum between 10^{-2} and 10^{-4} Pa.

As Table 5.1 shows, the vapor pressure of ice is almost 0 at 183 K. At this temperature, only a strong atmospheric depression applied on the specimen will be capable of extracting the water. Under these conditions, the sublimation of the ice in very small samples (<0.5 mm^3) takes several days. A particularly high vacuum is required for this operation, as is an adjustable thermostatic enclosure.

Table 5.1 Ice vapor pressure as a function of temperature

Temperature (K)	Ice vapor pressure (mmHg)
183	0.000070
203	0.00194
213	0.00808
233	0.0966
253	0.776
263	1.950
273	4.579

6 Actions Resulting in a Change in Material Properties

Certain techniques require the use of chemical compounds intended to stabilize the biological material. They create chemical bonds between the biological material and the additional chemical compound that definitively modify the chemical composition of the material. This is the case for chemical fixation and "positive-staining" contrast.

Other techniques result in changes in the material's mechanical behavior during preparation. This is the case for infiltration and embedding. They require the use of polymer resins that will be intimately bound to the sample, around it in the case of embedding or infiltration. For hydrated samples, this step is preceded by dehydration if embedding takes place in non-polar resins.

6.1 Chemical Fixation Principles

Living matter can be considered as a very hydrated protein gel in which ions, sugars, lipids, amino acids, nucleic acids, etc. constantly move.

Chemical fixation consists of bridging the proteins, so as to create an artificial cross-linking of the gel by forming long insoluble chains in order to eliminate the water, while preserving the original structural network. What follows is a denaturing of the proteins and a loss of all of the small diffusible elements. By denaturing the proteins, chemical fixation blocks the enzymatic systems, preventing any subsequent deterioration of the specimen through autolysis.

The fixatives used are aldehydes (paraformaldehyde and glutaraldehyde) or powerful oxidants (osmium tetroxide, ruthenium tetroxide, etc.). The fixative binds to the protein by an addition reaction at the level of the double bonds or alcohol, aldehydes, or hydroxyl groupings.

Glutaraldehyde is a dialdehyde with the formula $OHC-(CH_2)_3\,COH$. It can react twice with the H^+'s and create bridges between the molecules.

Paraformaldehyde is a monoaldehyde of an indefinite formula $(CH_2O)_n$ (n = 8–100), because its degree of polymerization is variable, explaining the n index.

Aldehydes are good fixatives for proteins and nucleic acids. Glutaraldehyde can result in artificial bridges between proteins and free amine groups through the addition phenomenon. It also causes a change in protein spatial conformation, which could prevent the detection of enzymatic sites. This is why the use of paraformaldehyde is preferred as a fixative in immunocytochemistry. Paraformaldehyde has just one aldehyde ending. Furthermore, because of this it establishes unstable bonds with proteins, and the paraformaldehyde fixative mix is often supplemented with a weak concentration of glutaraldehyde. Aldehydes do not fix lipids or phospholipids, which compose the plasmic membrane.

Osmium tetroxide is a powerful oxidant due to its four oxygens (Fig. 5.18); it reacts with the double bonds situated between the two carbons, C=C.

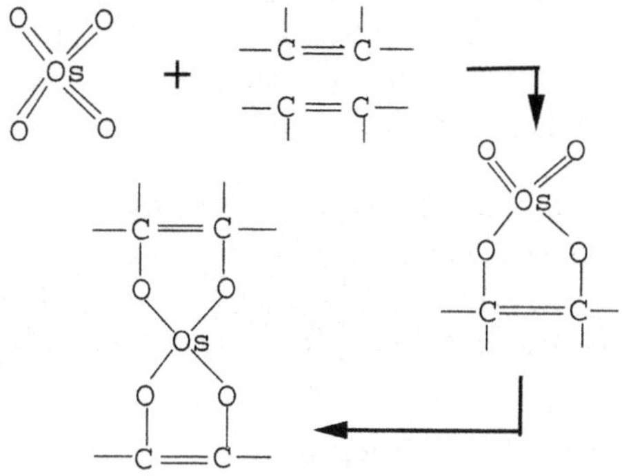

Fig. 5.18 Diagram of osmium tetroxide

Osmium tetroxide is a fixing agent for proteins and amino acids; it reacts partially with nucleic acids. It strongly fixes itself in compounds containing double bonds and in particular unsaturated lipids such as the triglycerides and phospholipids of membranes.

This last action is represented by the images of unbroken lines that are observed on chemically fixed membranes.

Since osmium tetroxide has a strong affinity for double bonds, it can split up molecules. As it is more soluble in alcohol than in water, it risks dissolving in alcohol during dehydration, causing ruptures in the long protein chains that become soluble. This phenomenon is called overfixing.

Osmium has the advantage of being a chemical element with a high Z. It increases the contrast of the preparation. Taking into account the scale of observation in TEM (on the order of a nanometer), it is essential to respect the physical–chemical characteristics of cellular fluids as much as possible to minimize the movements of water or ions in the tissue and the cells before cross-linking. This is why the fixative will be placed in solution in an appropriate "vector liquid" that aims to reproduce the *milieu intérieur* of the cells. This entire ensemble constitutes a fixative mixture.

The physical–chemical characteristics taken into consideration for balancing the vector liquid are pH, molecular concentration, and ionic concentration.

6.1.1 Constancy of pH

In an aqueous solution, certain fixatives are acidic (pH from 3.5 to 5) or alkaline (pH around 10). Additionally, during fixing, there is a release of H^+ ions coming from the above-mentioned chemical reactions, causing acidification of the fixative mixture. A buffer system is used to trap these H^+ ions. This buffer system is composed of ionized substances that contain two salts, one salt and one acid, or one salt and one base. The buffer system chosen must meet certain criteria:

- Efficacy in the selected area, depending on pH
- Good water solubility
- Good penetration in biological systems
- Reduced side effects. In particular, the toxic effects on cells must be prevented, as well as precipitations due to the presence of some cations (in particular with Ca^{2+}). The ionic equilibrium of the fixative in relation to the ionic balance of the cell must also be monitored

The most commonly used buffers are mineral buffers (phosphates or cacodylate) or organic buffers (s-collidine, P.I.P.E.S).

The pH level is chosen depending on the *milieu intérieur* of the tissue (e.g., blood plasma). It is usually near neutral (7.2–7.4).

6.1.2 Molar Concentration

Differences in molar concentrations between the fixative mixture and the tissue lead to movements of water and ions across the plasma membrane, which is semi-permeable. The molar concentration of the fixative mixture is, as with pH, a replication of that of the *milieu intérieur* of the cells and tissues. For mammals, it is around 300 mOsm. The osmolarity correction is calculated for the entire fixative mixture. It has a component due to the fixative itself, a component due to the buffer system, and possibly an additional component (salt or sugar).

The semi-permeability of the plasmic membrane is not destroyed by the aldehydic fixatives, so the osmotic balance between the cell and the fixative must be watched until osmic fixation. It is necessary to rebalance the rinsing vector liquid after aldehydic fixation and compensate for the osmolarity of the fixative, which is no longer in the mixture.

6.1.3 Ionic Concentration

There are many movements of ions within the cells; they influence the electrical conduction between cells, muscular contractions, the entry and exit of molecules through the plasmic membrane, secretion, etc. Therefore, it is important to create a minimum of disturbances at this level. But the reference level is difficult to know, since natural ionic movements are extremely rapid. In the absence of precise data, it is necessary to be sure not to add too much of a single ion; therefore, mineral buffers containing both sodium ions and potassium ions, or better still, organic buffers, which are neutral from this standpoint, are used.

Furthermore, experience shows that the presence of divalent Ca^{2+} or Mg^{2+} cations in the vector liquid provides better membrane preservation.

From a preparation standpoint, the microstructure of biological materials will determine the fixation and inclusion protocol.

The different types of biological material microstructures are classified by an ascending degree of complexity, from materials that may be considered amorphous up to mixed–composite materials (Fig. 5.19).

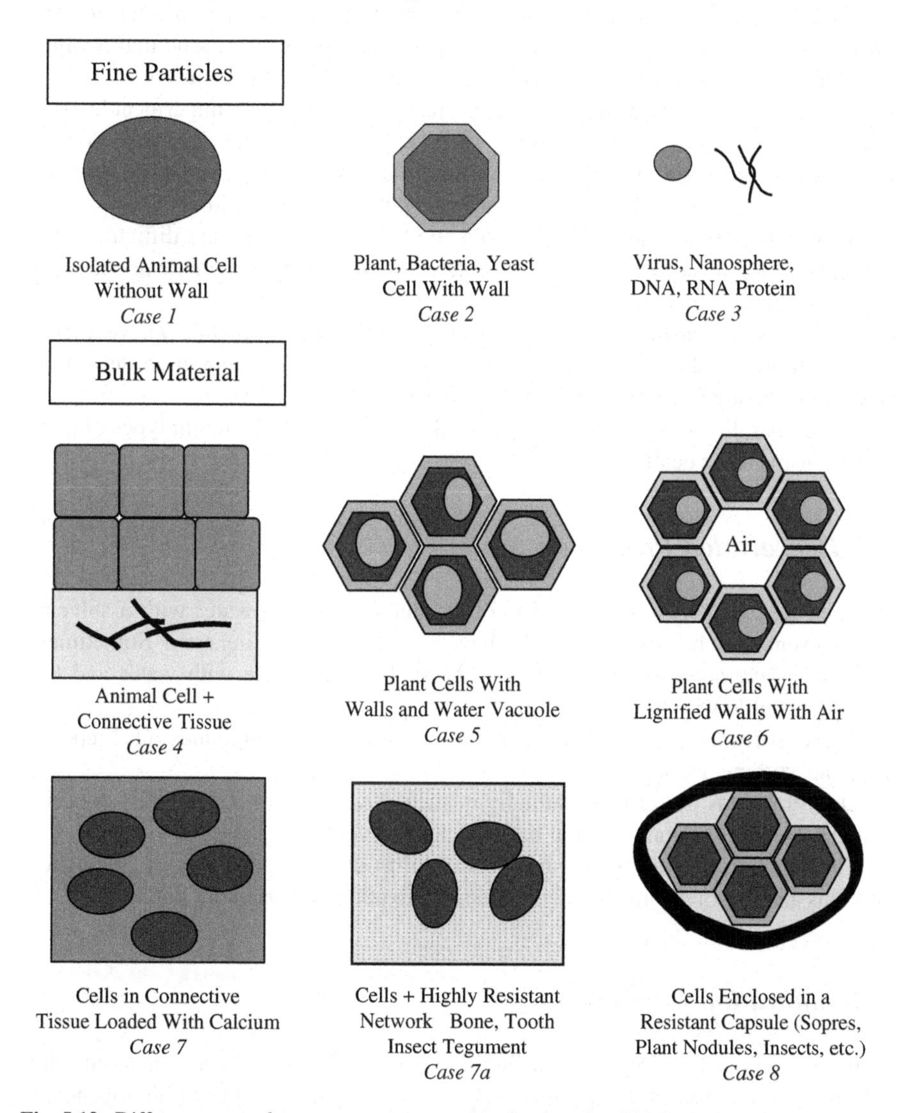

Fig. 5.19 Different types of structures encountered in the domain of biological materials

What differentiates the structure is hardness heterogeneity, component brittleness, the presence of pores filled with air or water, and the difficulty of liquid penetration.

Isolated cells or monolayer cells, as well as isolated organelles (mitochondria, DNA, RNA, etc.) and viruses are very easy to fix (cases 1 and 3).

Bacteria and yeasts, as well as certain protozoa, are unicellular organisms, endowed with walls that make penetration by liquids more difficult (case 2). It is often necessary to dissolve these membranes using enzymatic digestion.

Cells are usually combined with one another to form a tissue, and often several tissues are combined to form an organ (case 4). The connective tissue that is often associated is a water trap that will require increased dehydration time.

Plant cells are surrounded with a relatively complex and impermeable wall (case 5). These walls are often poorly permeable and necessitate allowing longer action times than for animal cells. Animal and plant cells can also contain air (case 6). Air must be replaced under vacuum by the fixation liquid.

Support tissues such as cartilage, bone, tooth, cuticle will present difficulties for the embedding. It is often necessary to soften these tissues by enzymatic or chemical digestion (cases 7 and 7a).

Very small organisms (worms, insects, parasites, and plant nodules) have impermeable, rigid membranes in which one must make openings using microdissection in order to ensure penetration of the preparation liquids (case 8).

The preparation protocol must be adapted each time to the different types of cells or organisms to be dealt with.

6.2 Dehydration Principles

As seen earlier, dehydration consists of gradually replacing water with a solvent. This solvent must be miscible with both water and the embedding resin. Sometimes, two successive solvents must be used, the first one miscible with water and the second miscible with the first and the resin.

These solvents are alcohols, acetone, or aromatic compounds (1,2 epoxy, propane, styrene, etc.).

Dehydration is performed gradually in successive baths of increasingly concentrated solvents; it is completed in baths of pure solvent.

The solvent (alcohol or acetone) is harmful to organic molecules. It coagulates proteins and dissolves lipids, unless chemical fixation is carried out beforehand.

6.3 Infiltration Principles

Infiltration consists of replacing a liquid or gaseous phase within a material with a liquid polymer that will then be hardened to produce a sample of homogeneous hardness.

In the case of porous materials, resin will replace the air. If the porosity is coarse, the air is replaced directly with resin under a primary vacuum. If the porosity is fine, the air is replaced under vacuum with a solvent or water and then substitution baths are used as in the case of hydrated samples.

For materials in liquid solution, infiltration follows dehydration. The liquid or gaseous phase of the sample is gradually replaced by a solvent and resin mixture. Since it is difficult to make a more or less viscous liquid penetrate cavities or porosities, several mixing baths are carried out with an increasing ratio of resin to solvent. This has the advantage of decreasing resin viscosity and facilitating penetration. The proportions of the mixture and the time required between each bath depend on several factors: the type of resin, its viscosity, the size and density of the material, and the size of the "cavities" the resin must penetrate. To ensure proper penetration, the volume of the solvent/resin solution must be at least 10 times greater than the sample volume.

In the final phase, infiltration under a primary vacuum has the advantage of quickly eliminating the solvent or the gaseous phase, while promoting penetration of the resin into the material. This also helps to degas the resin, as air may have been introduced during preparation of the mixture. Slow lateral or rotary agitation also improves penetration while reducing the time required between each bath.

The last step in the infiltration procedure corresponds to the embedding of the sample.

Infiltration (and embedding) polymers are mixtures in the liquid state, more or less viscous, which hardens through polymerization. Polymerization results from chemical bonds that form between the monomer and one or more hardeners resulting in an intertwining (cross-linking) of the molecules forming a solid network. Once polymerization occurs, the polymer is of a variable hardness, depending on the type of polymer or the percentage of the compound mixture.

The infiltration resin is selected based on several parameters essential to the preparation and observation of samples in the TEM:

- There must not be any shrinkage (reduction of volume) or expansion of the resin following polymerization; this would result in either a resin/sample separation during preparation or a change in the morphology of the material.
- The hardness of the resin must be as close as possible to the hardness of the material, so that the polymer's response to the different types of preparation mechanisms is identical or similar to the material.
- The polymerized resin must be stable under the electron beam.
- It must not present texture under observation.
- In the case of infiltration, the resin mixture must, in the liquid state, be of low viscosity in order to properly penetrate the samples.

There are many infiltration resins available, but here we will concentrate only on those most commonly used for preparation in electron microscopy: epoxy and acrylic resins.

Epoxy resins are generally composed of one or more monomers, one or more hardeners, a reaction accelerator, and sometimes a plasticizer or a flexibilizer that acts as a hardener or a plasticizer. Epoxies are generally chemical substances containing oxygen bridged onto a carbon–carbon bond. They are also referred to as oxacyclopropanes (systematic nomenclature) or also to oxiranes. Epoxy resins are polyaryl ethers of glycerol with a terminal epoxy group. They can be polymerized using a wide variety of hardening agents, which are anhydrous aromatics that, through a chemical reaction, are added to the epoxy groups in order to form a 3D structure. The macromolecules intertwine and form a solid network. These resins are said to be cross-linked. The relative proportion of each component (monomers and hardeners) can produce resins of different hardnesses. The reaction is triggered by the addition (in a very small quantity) of an accelerator, often from the family of aliphatic polyamines. Of all the embedding resins, epoxy resins provide the smallest shrinkage rates (2% maximum). Polymerization is uniform. They are non-polar and only polymerize in the presence of traces of water. Their adhesion to most materials is excellent.

They are polymerizable under heat (from 333 to 393 K). Their polymerization time is relatively long (from a few hours to several days).

Acrylic resins consist of autopolymerization components that are stabilized in their commercial form. At the moment they are used, a destabilizer is added (benzoyl peroxide) to trigger polymerization. Acrylic resins are usually polar and polymerizable under heat or UV rays, at room temperature or to 277 K. Some (Lowicryl resins) polymerize at low temperatures, from 203 to 273 K under UV radiation. For these resins, a photoinitiator must also be added at very low temperatures. Acrylic resin hardens, is thermoplastic, and resists chemicals. Methacrylate resins are composed of one or more methacrylates, such as methyl methacrylate, butyl methacrylate, hydroxyethyl methacrylate (HEMA), a plasticizer, and a destabilizer (benzoyl peroxide). Generally speaking, acrylic resins in the liquid state are more fluid than epoxy resins. Their ability to polymerize under cold is ideal for infiltrating or embedding heat-sensitive materials. The most commonly used resins and their properties are listed in Table 5.2.

6.4 Embedding or Inclusion Principles

The purpose of embedding a material (also called inclusion), usually one with small dimensions, using a polymer (a resin) is to enable the material to be manipulated and/or to protect its surface from possible damage during the preparation process. Embedding is used alone, i.e., without infiltration, in materials science for non-porous materials and in life sciences for samples with very small volumes (single-layered cells, for example). Regardless of whether they belong to materials science or life science, porous or hydrated materials must be infiltrated beforehand. Biological materials are always fixated beforehand, except for organic matrices treated as minerals (bone, teeth, etc.). In this case, the embedding resin is the same as that used for infiltration. Generally speaking, the types of embedding resins are the same as those used for infiltration.

The sample must present an absolutely clean contact surface free of contaminants and grease or any other dirt which could prevent the adhesion of the polymer around the sample. The process should not generate a chemical reaction between the polymer and the material.

Table 5.2 Properties of the most commonly used infiltration–embedding or embedding resins

Resin	Low fluidity (+) to very fluid (+++)	Solubility	Water miscibility	Stability under electrons	Ease of cutting	Polymerization
Epoxy: (Araldite/ Epon)	+	Alcohol, acetone, epoxy 1,2 propane	Non-polar	Very good	+++	Under heat
Epoxy: ERL (Spurr)	++	Alcohol, acetone	Non-polar	Good	++	Exactly 333 K
Epoxy: Epofix	++		Non-polar	Good	++	Room temperature
Acrylic: LR white	+++	Water, alcohol, acetone	Polar	Good	++	At 333 K or UV at 277 K
Acrylics: Lowicryl K4M, K11M	++	Water, alcohol, acetone	Polar	Average	+	<273 K UV
Lowicryl: HM20, HM22	++	Alcohol, acetone	Non-polar	Average	++	<273 K UV

The resin must be poured (around the sample) in an embedding mold whose chemical composition does not interact with the components of the embedding polymer.

The time and method of resin polymerization are specific to the type of embedding resin used and are provided by the manufacturer.

To improve adherence of the resin to the surface of the material, a preliminary treatment (primer) composed of 1–2% silane mixed with an ethanol solution may be used.

For materials presenting a high surface roughness, the use of a procedure similar to that used for infiltration (successive baths), as well as embedding under vacuum, could improve adhesion of the polymer to the material surface.

6.5 "Positive-Staining" Contrast Principles

Biological materials and polymers often present poor contrast during TEM observation. Since image contrast results from variations of electron density within the thin slice of the material, the principle behind this technique is to make heavy metals react with the structures to be observed in order to stop as large a number of electrons as possible. The addition of this contrast reagent must be selective, i.e., it must act on certain structures and not on others to be able to differentiate them. This is particularly evident in a mixture of two polymers where one will be stained and the other not, helping to locate the different phases of these two components in the material.

This contrast is said to be positive in order to differentiate it from another so-called negative contrast (to be discussed later). Positive staining presumes a stable link between the contrast reagent and the structures. Rinsing eliminates excess contrast reagent that has not reacted.

The metals used as contrast reagents are osmium (Os), ruthenium (Ru), lead (Pb), uranium (Ur), tungsten (W), silver (Ag), etc. They come in the form of an oxide (Ru or Os tetroxide), salt (U or Pb), or acid (APT). Other forms of metal open double bonds or bind to CH, COH, OH, H, NH_2, etc., endings, causing a bridging chemical reaction, or they bind to one another through preferential physical adsorption. In this last case, the mechanism of action is poorly known and a specific reaction cannot be discussed.

For some polymers, the reaction is not produced directly. An agent that promotes the reaction can be used, e.g., formaldehyde, hydrazine, hydroxylamine, or other amines that will create the possibility of bonds with the contrast reagent. These agents change one of the compounds in order to render it reactive.

Sometimes heat or alkaline saponification can be used to produce the same effect.

Lead and/or uranyl is used on biological materials under strict pH and dissolution conditions. The mechanism behind the reaction is poorly known and does not appear selective; it appears that it is favored by the prior presence of osmium, for which these contrast reagents have an affinity.

In the case of this technique, the sample is chemically modified. Chemical analysis of the sample must take this into account.

7 Physical Actions Resulting in Deposition

7.1 *Physical Deposition*

Physical depositions (or coatings) are used in many techniques for preparing thin slices, as well as during preliminary preparation for obtaining continuous or holey support films. Deposition always consists of depositing carbon, metal, or a polymer, under a vacuum, in the form of very fine particles that remain individualized or form a continuous film.

When a fine particle sample does not present enough contrast for observation, the contrast is enhanced through shadowing techniques using metallic particles (platinum, chrome, etc.). If it is a crystal-plane sample presenting very small differences in contrast, decoration techniques involving the deposition of polymers or gold particles are used to highlight these different levels.

When investigating samples that cannot be introduced into the TEM, replica techniques are used to investigate the topography on replicas (prints) made of thin carbon films with or without the metallic shadowing effect.

Films are made to support fine particles and brittle materials, or those too small to be self-supporting. These films are generally carbon, polymer, or mixed polymer–carbon films.

An increase in contrast can be obtained using the "negative-staining" contrast technique. In this case, the coating is obtained using a chemical solution of one of the metals used previously.

7.2 Physics of the Coating Process

A continuous thin film on the surface of a substrate or a sample is obtained by the deposition of metal or carbon source heated until the equilibrium vapor pressure is reached. The transport of metal or carbon particles to the sample occurs through an appropriate vacuum.

The quality of the various coatings depends on the level of the vacuum and the possible presence of impurities causing contamination. Condensation of these particles on the sample will gradually form a solid continuous film. The quality of the film depends on the cleanliness and roughness of the sample surface, as well as the film grows mechanism.

The formation of a continuous thin film results from two simultaneous processes: nucleation and growth. The first particles arrive on the substrate and are adsorbed by its surface (Fig. 5.20a). They are not in thermal equilibrium with the substrate and they migrate to its surface (Fig. 5.20b). This migration is accompanied by collisions and a recombination of atoms resulting in the formation of clusters (Fig. 5.20c), which grow with the arrival of new particles until islets are formed: this is called nucleation (Fig. 5.20d). The growth of the islets continues (Fig. 5.20e). Figure 5.20f shows the shape of the islets in cross section during the deposition process. Then the islets pile up on top of one another by means of coalescence (Fig. 5.20 g) until they constitute a continuous thin film covering the entire surface of the substrate (Fig. 5.20 h). Figure 5.20f shows the shape of the islets in cross section during the deposition process. If evaporation is stopped before a thin film is obtained, the size of the particles visible on the sample will be that of an aggregate of atoms.

Usually, the particles that constitute the thin layer are randomly oriented. For metallic films, the size of the particles influences their organization: If the particles are smaller than 2 nm, the layers have an amorphous structure, if their dimension is greater than 2 nm, the film is polycrystalline. There is, however, an exception: If the substrate is a thin film, the thin film is formed of particles oriented parallel to one another and connected to one another by low-angle grain boundaries. These are called epitaxial layers.

In the applications with which we are concerned, either a thin continuous film or dispersed particles are deposited for the shading method. In both cases, the intent is to obtain a very small particle size so that they can minimally disturb observation. This size is between approximately 0.5 and 2 nm. Here we will deal with the procedures most commonly used by microscopists.

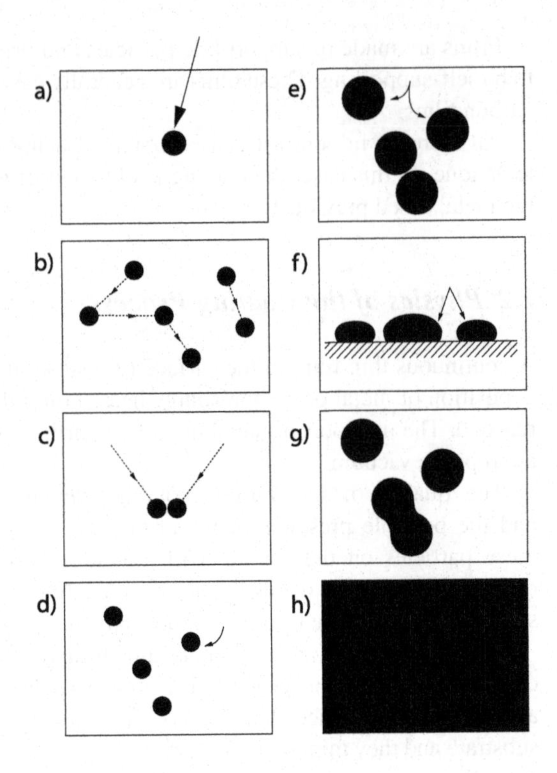

Fig. 5.20 Formation stages of the continuous thin film: (**a**) arrival of the first atom on the substrate; (**b**) migration and re-evaporation; (**c**) collision and recombination of atoms; (**d**) nucleation; (**e**) growth; (**f**) form of cross-sectional islets; (**g**) coalescence; and (**h**) film continuity

Several parameters are involved in particle size:

(A) The nature of chemical elements used as sources
(B) The different particle-production methods

1. Evaporation by the Joule effect
2. Evaporation using an electron gun
3. Pulverization using an ion gun
4. Cathodic pulverization

(C) The quality of the vacuum acting on the transport of particles during their deposit
(D) The nature, surface state, and temperature of the substrate during the coating process

7.2.1 Nature of Chemical Elements Used as Sources

Coatings are generally composed of atoms of carbon, carbon–metal, metal, or polymer molecules. The atoms produce finer grains than the molecules. The metals most commonly used are platinum, chrome, tungsten, and osmium, as the grains obtained are very small (unlike gold grains, which coalesce more quickly). Each metal has different evaporation constraints (evaporation temperature, vacuum quality, toxicity, etc.), which must be taken into account when choosing the technique.

7.2.2 Different Methods of Particle Production

In all of the evaporation methods, the thickness and quality of the deposit essentially depends on the following:

(a) The distance between the source and the substrate
(b) The voltage and intensity of the current

Evaporation by the Joule Effect

Joule-effect evaporation is performed in order to obtain a metallic layer or an amorphous carbon layer (Fig. 5.21).

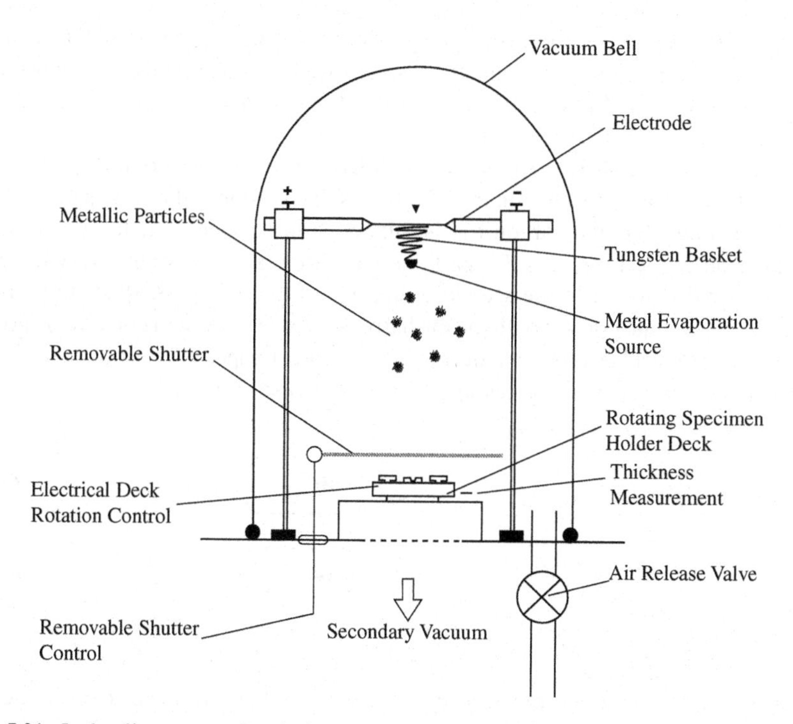

Fig. 5.21 Joule effect evaporation device

The setup used comprises a vacuum chamber connected to a pumping system for producing a vacuum of 10^{-4} Pa, in which is placed a pair of electrodes to evaporate the metal, a specimen holder, and a removable shutter. The stage can be tilted and rotated.

Generally, an oscillating quartz microbalance is placed near the object, providing real-time measurement of layer thickness. Another way is to put a filter paper close to the sample. Part of the filter paper is protected from evaporation to allow seeing the coloring changes, by comparing the grey color of the two filter paper areas. The gray coloring changes depend on the coating thickness. This system is used to determine approximate layer thickness.

The metal evaporation source, generally a wire with a diameter of approximately 1 mm, is located in a tungsten basket placed between the two electrodes. When the chosen vacuum is reached, the setup is heated until evaporation begins.

The evaporated metallic particles are projected into all directions, particularly downward, where the stage holds the object. Evaporation is very rapid. Once the desired thickness is reached, monitored by the in situ thickness control, the heat is suddenly shut off and the object can be removed from the chamber once air is allowed inside.

The evaporation temperatures of different metals used are found in specialized publications and are essentially dependent upon the pressures in the vacuum chambers.

There are many other source-support devices. In particular, cradles that require inverting the positioning of the substrate, since evaporation can only occur upward. Also in the specialized publications (L. Holland) are tables with different setups to be used and their conflicts with different metals.

For the evaporation of carbon by the Joule effect, the evaporation device is replaced by two carbon rods. Figure 5.22 shows two carbon-rod device setups. The first setup shows that the carbon rod is thinner in the center so as to prevent less resistance during heating. It is this part that will evaporate first when a very intense current goes through it. In the second setup, one of the rods is sharpened to a fine point and placed into contact with the unsharpened rod. The system contains a spring that ensures contact between both during evaporation. Evaporation is obtained by an electrical arc generated by the passage of a very intense current.

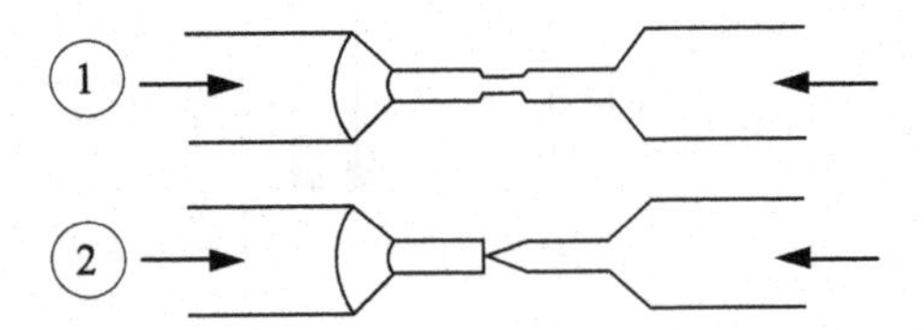

Fig. 5.22 Device for carbon evaporation by the Joule effect

The carbon rods can be replaced by a carbon braid. This system is very useful for obtaining good reproducibility.

Evaporation Using an Electron Gun

Evaporation by means of electron bombardment is used to obtain metallic coatings with very fine grain sizes.

The setup used is represented in Fig. 5.23. The anode is composed of a graphite pencil that has been hollowed out at one end. A disk of the metal to be evaporated (platinum, chrome, etc.) is then placed in this hollow. This electrode is placed within an electrical field created by a tungsten filament, corresponding to the cathode. The system of shutters makes it possible to channel the flow of particles to the specimen to obtain a better output. The setup is in the upper part of the chamber, which is placed under a high vacuum of 10^{-4}–10^{-5} Pa. The specimen is placed on a specimen holder stage that can sometimes be tilted and rotated some tens of rotations per minute. It is then placed under the gun at a short distance (5–10 cm). This system makes it possible to evaporate the metal with a weaker current and much less heating than previously, which allows much smaller particles to be obtained.

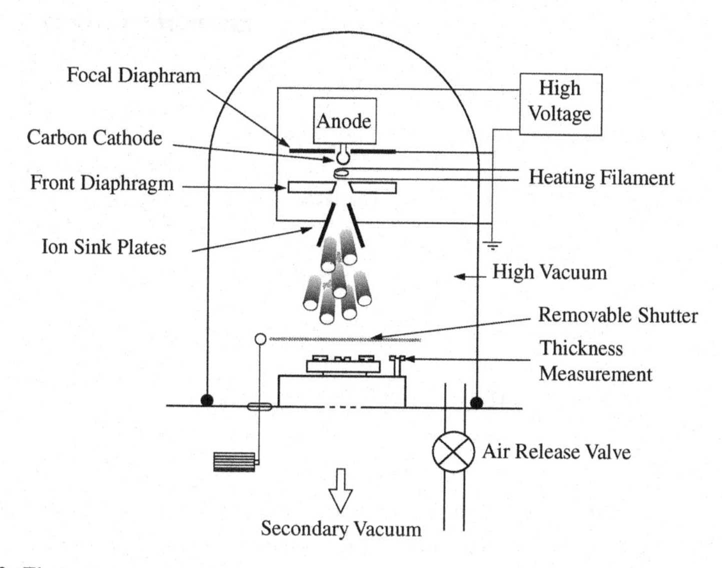

Fig. 5.23 Electron-gun evaporator

Pulverization Using an Ion Gun

In this device, an ion gun is added to the evaporation source, so that the coating is made under the ion beam, in what is called ion beam-assisted deposition (IBAD).

This system is relatively recent and designed to overcome the often poor quality of coatings made by simple evaporation (bad aging, poor density, poor adherence, etc.). The setup used is represented in Fig. 5.24 and shows a classic evaporator equipped with an ion gun. Coatings are made in a chamber under a vacuum of 10^{-5}–10^{-6} Pa. The deposition technique consists of evaporating the material present within a crucible while simultaneously bombarding the film during its growth, using a beam of energetic ions coming from a plasma source. The energy provided by the ions is transferred to the atoms derived from the evaporation, which are adsorbed onto the surface of the substrate. Depending on the type of atoms evaporated and the plasma-bombardment ions, chemical reactions may occur, giving rise to defined

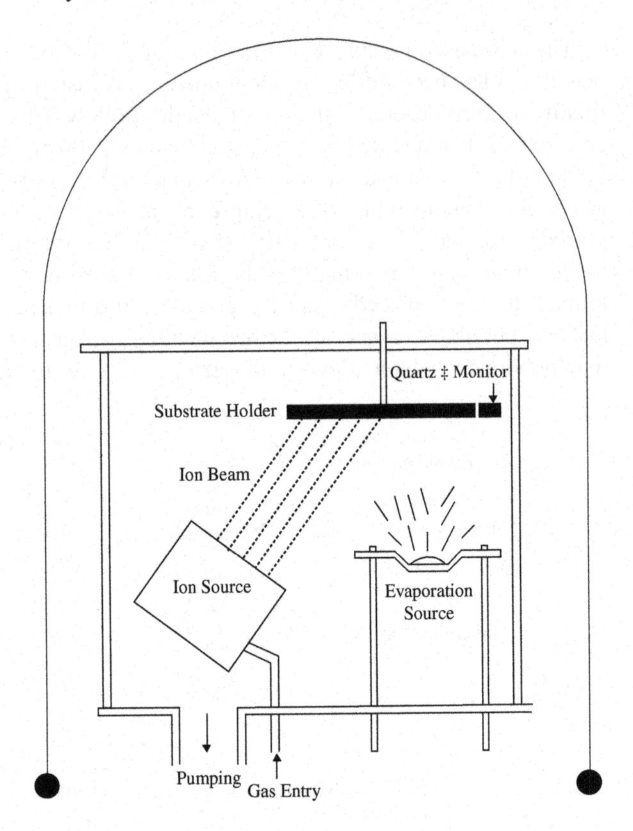

Fig. 5.24 Ion Gun Evaporator device

components. The advantage of this technique is that a coating can be made at room temperature, or any other temperature, if a device is included for heating or cooling the sample. Therefore, this technique is used to synthesize a multitude of materials; multilayer structures can be obtained using simultaneous and localized evaporation of several materials by moving the substrate. Since there is no contact between the plasma and the growing layer, contamination is minimal. One of the purposes of this technique is to eliminate the columnar structure of deposits that appears during their growth. Ion bombardment significantly influences the arrangement of the atoms of the layer being formed; adherence to the substrate is improved, internal stresses are reduced, and impurities disappear. The optical characteristics of the layer as well as its electrical resistance are markedly improved.

Cathodic Pulverization

This system is widely used for scanning electron microscopy. It consists of ejecting atoms from a metallic cathode using an ionized plasma gas, usually argon, under a residual vacuum on the order of a few tens of pascals. These ions are deposited "in a rain" on the specimen placed on the anode.

In TEM, this technique is only used to deposit polycrystalline gold particles on the surface of a carbon TEM support film. These gold particles help in the calibration of diffraction diagrams.

7.2.3 Vacuum

The size of the grains deposited depends on the quality of the chamber vacuum and the type of plasma. Classic pumping contains a roughing pump combined with an oil diffusion pump that creates a vacuum on the order of 10^{-4} Pa.

This vacuum can contain the vapors of residual oil coming from the pump that contaminate the chamber and especially the surface of the specimen during deposition. Contamination can be reduced by placing a liquid-nitrogen trap above the diffusion pump. Pumping using a turbomolecular pump does not have this drawback, since there is no heated oil in this system.

Pumping begins at an atmospheric pressure and continues until a pressure of 10^{-4} Pa is reached with the same pump.

This is the ideal system, but it can be used only in small-volume chambers, otherwise pumping would take too long.

Cryo-pumping is used for some freeze-fracture devices. It is the cleanest vacuum, allowing acquisition of pressures down to 10^{-4} Pa. The ionic vacuum can also be used if a very high vacuum from 10^{-5} to 10^{-6} Pa is needed.

7.2.4 Substrate

The surface of the substrate or the specimen must be very clean. The increase in substrate temperature during deposition increases the kinetic energy and therefore the mobility of the particles. The result is increased particle size. To prevent this, specimen holders cooled with either water or liquid nitrogen can be used.

The crystalline state of the substrate surface can be used wisely to make epitaxial coatings.

For fine particles, the dispersion liquid can contain wetting agents, salts, or other compounds that impede the proper distribution of metallic particles.

Eliminating them is recommended insofar as possible.

7.3 Techniques Involving a Physical Deposition: Continuous or Holey Thin Film, Contrast Enhancement by Shadowing or Decoration, Replicas, and Freeze Fracture

Continuous or perforated thin film: To obtain a thin and self-supporting carbon film, an NaCl single crystal is used as a deposition substrate. The gem salt must then be dissolved on the surface with water. A freshly cleaved sheet of mica can also be used. The film is unstuck by dunking/floating it on the surface of some water. The film is recovered on TEM support grids (see Volume "Techniques", Chapter 2 , Sections 13 and 14).

To deposit carbon on support grids coated with a thin polymer film (continuous or holey), the classic Joule-effect evaporation techniques are used.

Contrast enhancement using shadowing or decoration is essentially used for fine particle materials. To obtain a shadowing effect on the sample, an evaporation angle of a few degrees with regard to the plane of the thin slice must be used. The angle will be lower if the relief of the structure to be highlighted is low. This requires a special setup in the evaporation chamber. Generally, two evaporation heads are used. The first is used to carry out low-angle evaporation of the metal in order to produce a shadowing effect and the second is used to reinforce the film obtained by a layer of amorphous carbon evaporated perpendicular to the surface. The use of masks enables two successive evaporations without disrupting the vacuum in the chamber. The specimen holder must be tiltable from the outside.

The decoration technique uses either evaporation by the Joule effect of a polymer or evaporation of gold. This latter technique highlights the intrinsic crystal organization of the object. The deposited atoms will select preferential sites and provide indications of the surface state of the material.

7.3.1 Replica Techniques

Direct replica: The direct-replica technique requires the use of the heavy metal shadowing technique followed by a carbon coating in order to form a continuous film, constituting the print. In this case, the object is dissolved and the microscopic investigation is conducted on this print.

Indirect replica: In this case, the initial object is replaced by a polymer mold, which is then shadowed. A carbon film is deposited on it as in the direct-replica technique. The polymer mold is dissolved and TEM observation is carried out on the print of the mold.

Extractive replica: The extractive replica technique uses the evaporation of a carbon film deposited onto the surface of the object in order to enable the extraction of particles found there, enabling their observation in the microscope.

Freeze fracture: This is also a replica technique, but instead it is applied to the fractured face of an object. It applies to bulk, soft, and hydrated materials or fine particle materials in liquid solution. It is conducted on a frozen sample. After fracturing, and while the object is still cold, the shadowing technique is applied, followed by the evaporation of a carbon film. Investigation under the microscope is carried out on the print obtained after the sample is dissolved.

7.3.2 Contrast Enhancement by Physical Coating: "Negative-Staining" Contrast

In the case of the negative-staining contrast enhancement technique, the stain will not react chemically with the material, but create a dark background on which the structure appears bright, revealing its form, dimensions, and external ornamentation. This staining is not followed by any washing, which would totally eliminate the stain. This is a purely physical reaction of the adsorption of a chemical compound (the stain) on the support film. The stain in solution is ionized and binds to the support film or the sample surface because of its electrical charges. The intensity of this contrast will depend on the abundance of electrical charges on the sample surface and the surface of the support film. It also depends on the spatial size of the staining ion. For a material in aqueous solution, the stain penetrates the polar regions of the sample and replaces the water, helping to view very fine details on its surface. The stain is selected based on the sample and the structures to be highlighted. It is placed in solution with a specific pH that will induce its dissociation and therefore its electrical charges.

Negative staining has the advantage of protecting the sample from the drawbacks of desiccation. The stain evaporates faster than the sample and creates an envelope of stain around it, which maintains a certain degree of humidity.

There is a risk of crushing the sample on the support, increasing its dimensions. By chemically fixing the material beforehand and/or working with low electron doses, this phenomenon can be minimized.

The stains used are salts of heavy metals such as uranium, tungsten, and molybdenum. They are selected for their high water solubility in order to prevent the formation of crystals around the sample when drying. Their good stability under the electron beam depends on their boiling point, which preferably will be high. For tungsten ions obtained using phosphotungstic acid, the ion size is from 0,8 to 1,5 nm. For uranyl ions, the size is ranging from 0,4 to 0,5 nm.

Uranyl can be used in the form of an acetate, nitrate, or formate.

It is also used for "positive-staining" contrast. Note that in negative staining, a fraction of positive reactions may occur. To minimize this possibility, work is done at an acidic pH (less than 5); moreover, for a pH greater than 6, uranyl is unstable. Furthermore, uranyl has a known role as a fixative, on lipids in particular, but its mechanism is poorly understood.

Tungsten is most often used in the form of a potassium salt or sodium salt of phosphotungstic or silicotungstic acid.

Ammonium molybdate seems to perform better than the previous stains for revealing the structural details of membranes. It would also be less toxic and would reversibly inhibit metabolic activities (Fig. 5.25).

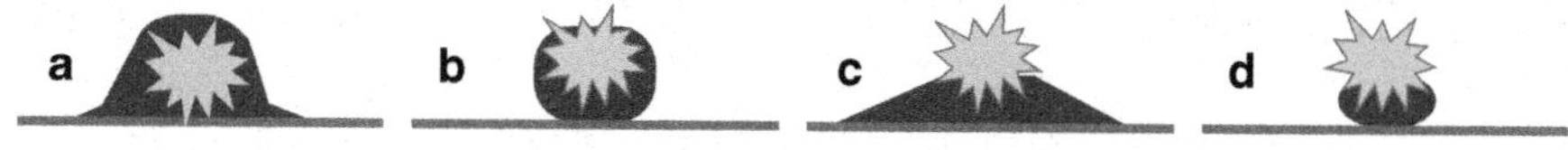

Fig. 5.25 Different stain-distribution possibilities based on the respective polarities of the support film and the sample surface: (**a**) polar sample and film, (**b**) polar sample and non-polar film, (**c**) non-polar sample and polar film, and (**d**) non-polar sample and film

Bibliography

Mechanical Action

Al-Amoudi, A., Studer, D., and Dubochet, J. (2005). Cutting artefacts and cutting process in vitreous sections for cryo-electron microscopy. *J. Struct. Biol.*, **150**(1), 109–121.

Dorlot, J.-M., Baîlon, J.-P., and Masounave, J. (1986). *Des matériaux, deuxième édition revue et augmentée*, Editions de l'École Polytechnique de Montréal.

Dupeux, M. (2004). *Aide-mémoire Science des matériaux, Sciences SUP*, Dunod.

Guinier, A. and Julien, R. *La matière à l'état solide*, Liaisons scientifiques Hachette.

Mercier, J.-P., Zambelli, G., and Kurtz, W. (1987). *Introduction à la Science des Matériaux*, Presses Polytechniques et Universitaires Romandes.

Reid, N. and Beesley, J.E. (1991). *Sectioning and Cryosectioning for Electron Microscopy in Practical Methods in Electron Microscopy*, vol. 13. (ed. A.M. Glauert). Elsevier, Amsterdam, 1–245.

Chemical Action

Goodhew, P.J. (1972). *Specimen Preparation in Materials Science, Practical Methods in Electron Microscopy*, Audrey M. Glauert.

Morel, G. (1995). *Visualization of Nucleic Acids*, Ed. CRC Press, Boca Raton, London, Tokyo.

Shigolev, P.V. (1974). *Electrolytic and Chemical Polishing of Metals*, Freund Publishing House.

Ionic Action

Giannuzzi, L.A., Prenitzer, B.I., Drown-MacDonald, J.-L., Shofner, T.L., Brown, S.R., Irwin, R.B., and Stevie, F.A. (1999). Electron microscopy sample preparation for the biological and physical sciences using focused ion beams. *J. Process Anal. Chem.*, **IV**(3, 4), 162–167.

Obst, M., Gasser, P., Mavrocordatos, D., and Dittrich, M. (2005). TEM-specimen preparation of cell/mineral interfaces by focused ion beam milling. *Am. Mineralo.*, **90**, 1270–1277.

Prenitzer, B.I, Giannuzzi, L.A, Brown, S.R, Shofner, T.L., and Stevie, F.A. (2003). The correlation between ion beam/material interactions and practical FIB specimen preparation. *Microsc. Microanal.*, **9**, 216–236.

Stevie, F.A., Giannuzzi, L.A., and Prenitzer, B.I. (2005). *Introduction to Focused Ion Beams: Instrumentation, Theory, Techniques, and Practice* (ed. L.A. Giannuzzi and F.A. Stevie).

Vieu, Ch., Gierak, J., and Manin, L. (2001). La Nanofabrication au Service de la Préparation d'Échantillons pour la Microscopie Electronique en Transmission, Atelier de préparation d'Échantillons pour la Caractérisation des Matériaux Nouveaux Multiphasés par Microscopie Electronique en Transmission, CSNSM Orsay, CNRS Formation.

Vieu, C., Gierak, J., Schneider, M., Ben Assayang, G., Marzin, J.-Y. (1998). Evidence of depth and lateral diffusion of defects during focused ion beam implantation. *J. Vac. Sci. Technol.*, **B16**, 1919.

Actions Resulting in a State Change of Materials Containing an Aqueous Phase

Burton, E.F. and Olivier, W.F. (1935). The crystal structure of ice at low temperature. *Proc. R. Soc. Lond.*, **153**, 166.

Brüggeller, P. and Mayer, E. (1980). Complete vitrification in pure liquid water and dilute aqueous solutions. *Nature*, **288**, 569.
Cavalier, A., Spehner, D., Humbel, B.M. (2008). *Handbook of Cryo-Preparation Methods for Electron Microscopy*. Editions CRC Press, Boca Rotan, FL.
Dubochet, J. and McDowall, A.W. (1981). Vitrification of pure water for electron microscopy. *J. Microsc.*, **124**, RP3-RP.
Morel, G. (1995). *Visualization of Nucleic Acids,* CRC Press, Boca Raton, London, Tokyo.
Richter, K., Marilley, D., and Dubochet, J. (1991). Cryo-electron microscopy of vitrified biological material. *Eur. J. Cell. Biol.*
Sawyer, L.C. and Grubb, D.T. (1996). *Polymer Microscopy*, 2nd edition. Chapman et Hall ed.
Steinbrecht, R.A. and Zierold, K. (1987). *Cryotechniques in Biological Electron Microscopy*, Springer, Berlin, 114–131.
Studer, D., Graber, W., Al-Amoudi, A., and Eggli, P. (2001). A new approach for cryofixation by high-pressure freezing. *J. Microsc.*, **203**, 285–294.

Actions Resulting in a Change in Material Properties

Carlemalm, E., Garavito, R.M., Acetarin, J.-D., and Kellenberger, E. (1985). *J. Microsc.*, **140**, 55–63.
Luft, J.H. (1961). Improvements in epoxy resin embedding methods. *J. Biophys. Biochem. Cytol.*, **9**, 409.

Physical Actions Resulting in a Deposit

Goldstein, J.I., Newbury, D., Joy, C., Lyman, C., Echlin, P., Lifshin, E., Sawyer, L., and Michael, J. (2003). *Scanning Electron Microscopy and X-Ray Microanalysis*. Kluwer Academic/Plenum Publishers, New York.
Hayat, M.A., Miller, S.E. (1990). *Negative Staining*. McGraw-Hill Publishing Company.
Holland, L. (1966). *Vacuum Deposition of Thin Films*, Chapman and Hall, London.
Richardt, A. and Richardt, I. (2000). *Les évaporations sous vide, théorie et pratique*, IN FINE.

[illegible] (1980) [illegible] in pure liquid [illegible] solutions. Nature 288:[illegible]
[illegible] (2006) Handbook of [illegible] Electron Microscopy [illegible] Edition. CRC Press [illegible], FL
[illegible] (2011) [illegible] of pure water [illegible] electron microscopy. [illegible] 124:[illegible]
[illegible] (1985) [illegible] Press, Boca Raton, London, [illegible]
[illegible] (1986) [illegible] Electron [illegible]
[illegible] material [illegible]
[illegible] (2000) [illegible] Microscopy, 2nd edition. Chapman & Hall, [illegible]
[illegible] (2008) [illegible]
[illegible] Biol. 124:[illegible]
[illegible] (2001) A new approach [illegible]
high-pressure [illegible] 203:[illegible]

Volume Rendering [illegible] Material Properties

[illegible]
[illegible]

[illegible]

[illegible]
[illegible]
[illegible]

Chapter 6
Artifacts in Transmission Electron Microscopy

1 Introduction

An artifact is damage caused by a preparation technique and can easily be confused with the sample's microstructure. Artifacts can be due to mechanical, chemical, ionic, or physical action. During TEM observation, especially in a TEM/STEM, other artifacts may be produced due to irradiation under the electron beam.

Therefore, there are two families of artifacts: those formed during the preliminary preparation steps and sample thinning and those formed under the effect of the electron beam during observation. Materials will be more or less sensitive to the formation of artifacts depending on their type and the nature of the chemical bonds.

2 Preparation-Induced Artifacts

Included among the possible artifacts are primary damage, formed by the preparation techniques, and secondary thermal damage, caused by a rise in the temperature during preparation or during observation under the electron beam. Table 6.1 shows all of the possible artifacts.

The artifacts and secondary thermal damage have been classified based on either the type of action (mechanical, ionic, chemical, and physical) involved in the preparation of the thin slices or their formation during the electronic radiation in the microscope.

The definition of each artifact is provided, as well as the preparation techniques from which they may be derived.

Lastly, this chapter presents illustrations of the artifacts most commonly encountered in the mechanical, ionic, chemical, and physical thinning techniques. It also provides illustrations of artifacts derived from the combination of several techniques in the fields of solid-state physics and biology.

J. Ayache et al., *Sample Preparation Handbook for Transmission Electron Microscopy*,
DOI 10.1007/978-0-387-98182-6_6, © Springer Science+Business Media, LLC 2010

Table 6.1 Table summarizing the various artifacts formed by the preparation techniques or during TEM observation

Mechanical preparation-induced artifacts	Ionic preparation-induced artifacts	Chemical preparation-induced artifacts	Physical preparation-induced artifacts	Artifacts induced during observation
Deformation	Redeposition	Material displacement	Deformation	Dehydration
Material displacement	Implantation	Selective dissolution	Microstructure change	Charge effects
Material tearing	Vacancies	Composition change	Segregation of liquid phases	Destruction
Cracks	Dislocation loops	Structural change		Contamination
Fractures	Cavities	Microstructure change		
Inclusion of abrasive grains	Fractures	Change of molecular bonds		
Dislocations	Roughness	Change of natural contrast		
Glide planes	Selective abrasion	Protein reticulation		
Twinning	Structural change	Residues		
Strain hardening	Microstructure change			
Selective abrasion				
Particle aggregation				
Roughness				
Structural change				
Microstructure change				
Crystal-network change				
Composition change				
Residues				
Secondary thermal damage				
Fusion	Fusion	Microstructure change	Deformation	Demixing
Phase transformation	Phase transformation	Changes in distribution in the phases	Particle aggregation	Amorphization
Loss of chemical elements	Loss of chemical elements		Fusion	Phase transformation
Amorphization	Amorphization		Phase transformation	Loss of chemical elements
	Demixing		Frost	Migration
				Fusion

2.1 Mechanical Preparation-Induced Artifacts

Deformation: Change in shape caused by the compression, stretching, or tearing of all or part of a sample's volume. This type of artifact can generate a change in matter conformation and distribution.

Techniques involved: sawing, mechanical polishing, dimpling, crushing, ultramicrotomy, cryo-ultramicrotomy, freeze fracture, and extractive replica.

Matter displacement: Transport by creeping or stretching of either matter on the sample surface or a particular phase in a multiphase material, caused by the friction of abrasive grains on the sample or by the cutting action.

Techniques involved: sawing, mechanical polishing, dimpling, tripod polishing, ultramicrotomy, and cryo-ultramicrotomy.

Tearing of matter: Localized loss of matter or of a particular phase in a multiphase material, caused either by the friction of abrasive grains on the sample or by the cutting action.

Techniques involved: sawing, mechanical polishing, dimpling, tripod polishing, ultramicrotomy, and cryo-ultramicrotomy.

Cracks: Crevices on the surface of the material, caused by the friction of abrasive grains on the sample, by the application of pressure (e.g., during cutting) or by thermal effects.

Techniques involved: sawing, mechanical polishing, dimpling, grinding, tripod polishing, ultramicrotomy, and cryo-ultramicrotomy.

Fractures: Splitting or separation of regions of the sample. The fracture can be caused by the rubbing of the sample on abrasive grains, the cutting action, the application of either pressure or tension, or thermal effects.

Techniques involved: sawing, mechanical polishing, dimpling, grinding, wedge cleavage, tripod polishing, ultramicrotomy, cryo-ultramicrotomy, and freeze fracture.

Inclusion of abrasive grains: The implantation by mechanical action of abrasive grains used during the technique, or even residues derived from contaminated polishing wheels.

Techniques involved: mechanical polishing, dimpling, and tripod polishing.

Dislocation: Linear structural defect (1D) resulting in an additional plane in an atomic stack. Dislocations are formed under the effect of a mechanical stress, e.g., impacts under the surface or tearing from abrasive grains on brittle materials.

Techniques involved: sawing, ultrasonic cutting, mechanical polishing, dimpling, grinding, wedge cleavage, tripod polishing, and ultramicrotomy.

Glide planes: Plane defect (2D) corresponding to the dense atomic plane of a crystal where dislocations propagate under the effect of a stress. This strong pressure on the crystal causes movements of whole atomic planes.

Techniques involved: sawing, ultrasonic cutting, mechanical polishing, dimpling, tripod polishing, and ultramicrotomy.

Particle aggragation: Agglomerate of particles formed by poor dispersion of isolated particles, or by migration of particles in solution during the dessication process. This is due to attractive forces between particles.

Techniques involved: fine particles dispersion technique.

Twinning: Plane defect (2D) corresponding to a joint or perfect mirror interface at the atomic scale. The atomic plane of the twin is common to the crystal networks of two grains. Their formation can be caused by a thermal phase transformation or mechanical stresses, a violent impact can cause twinning in brittle materials.

Techniques involved: sawing, ultrasonic cutting, mechanical polishing, dimpling, tripod polishing, and ultramicrotomy.

Strain hardening: Change in microstructure due to mechanical stresses, which induces a set of disturbances in the material involving dislocations and glide planes.

Techniques involved: sawing, ultrasonic cutting, mechanical polishing, dimpling, tripod polishing, and ultramicrotomy.

Selective abrasion: Thinning of matter at different rates by mechanical polishing. It can reveal a phase or lead to its disappearance through the faster abrasion of one phase over another. It can create surface roughness.

Techniques involved: mechanical polishing, dimpling, and tripod polishing.

Roughness: State of the non-plane surface revealing either scratches caused by material tearing or variations in thin slice thickness caused by selective abrasion or the tearing of matter.

Techniques involved: sawing, ultrasonic cutting, mechanical polishing, dimpling, tripod polishing, ultramicrotomy and cryo-ultramicrotomy.

Structural change: Partial or total change of the crystallographic organization caused by mechanical stresses. It can lead to a change in form, amorphization of the crystal network, a change in network, crystallization, or re-crystallization.

Techniques involved: mechanical polishing, dimpling, and tripod polishing.

Microstructural change: Partial or total change in the structural organization of microstructural components that can be caused by mechanical effects. It can result in changes in morphology, redistributions of phases, changes in crystal structure, precipitation, chemical gradients, formation of new phases, etc.

Techniques involved: mechanical polishing, dimpling, tripod polishing, and ultramicrotomy.

Crystal-network change: Change in the geometry of the crystalline structure and/or the crystalline motif. Any mechanical action can cause additional changes to the crystal network, such as dislocations, glide planes, twinning, and strain hardening.

Techniques involved: sawing, ultrasonic cutting, mechanical polishing, dimpling, tripod polishing, and ultramicrotomy.

Composition change: Change in the elemental composition of the sample due to a loss or dissolution of chemical elements during preparation.

Techniques involved: ultramicrotomy.

Residues: Deposits of matter extrinsic to the sample that may be polishing or sectioning residues resulting from preparation. They are superimposed on the thin slice and impede its observation. Residues are produced by mechanical actions during preliminary preparation or during final thinning.

Techniques involved: mechanical polishing, grinding, tripod polishing, ultramicrotomy, and cryo-ultramicrotomy.

2.1.1 Secondary Thermal Damage Induced During Mechanical Preparation

Fusion: Change of the solid–liquid physical state of a solid sample. It can be caused by an increase in temperature during mechanical preparation.

Techniques involved: mechanical polishing and tripod polishing.

Phase transformation: Total or partial transformation of the crystallographic and/or chemical structure, which can be caused by thermal effects during mechanical techniques.

Techniques involved: sawing, mechanical polishing, dimpling, and tripod polishing.

Loss of chemical elements: Selective loss of a chemical element, which may be caused by secondary thermal effects during mechanical preparation.

Techniques involved: sawing, mechanical polishing, dimpling, and tripod polishing.

Amorphization: Partial or total degradation of the chemical bonds of a crystalline sample, resulting in its destruction. This type of defect is caused by thermal effects during the mechanical action of preparation.

Techniques involved: mechanical polishing and tripod polishing.

2.2 Ionic Preparation-Induced Artifacts

Redeposition: Deposition of material previously milled by an ion beam, either intrinsic or extrinsic to the sample, which redeposits on the specimen surface during thinning. This is a form of contamination.

Techniques involved: ion milling and focused ion beam thinning (FIB).

Implantation: Accelerated ions from the beam penetrate the sample and remain trapped there, creating a new chemical type.

Techniques involved: ion milling and focused ion beam thinning (FIB).

Vacancies: Point defects (0D) corresponding to the loss of an atom in the crystal network. Accelerated ions can remove atoms or ions from the matrix, creating vacancies in the crystal network or creating defects (grain boundaries, twin boundaries, etc.). Accumulation of vacancies can form cavities.

Techniques involved: ion milling and focused ion beam thinning (FIB).

Dislocation loops: Partial dislocations that border an atomic-plane defect and close on themselves creating a dislocation loop.

Techniques involved: ion milling and focused ion beam thinning (FIB).

Cavities: Area free of matter inside a material. It is caused by the accumulation of vacancies and corresponds to volume defects (3D) that form an intra- or intergranular cavity.

Techniques involved: ion milling and focused ion beam thinning (FIB).

Fractures: Splitting or separation of regions of the sample. The fracture is caused by the relaxation of stresses due to the increase of temperature during ion thinning.

Techniques involved: sawing, mechanical polishing, dimpling, grinding, wedge cleavage, tripod polishing, cryo-ultramicrotomy, freeze fracture and focused ion beam thinning FIB.

Roughness: Variation of surface relief caused by selective abrasion.

Techniques involved: ion milling and focused ion beam thinning (FIB).

Selective abrasion: The ion beam can reveal phases if one of the phases abrades more slowly than the other. Selective abrasion can create an undulating surface, cause roughness by revealing dense atomic phases, or form peaks if there are precipitates or segregation zones (grain boundary, dislocation, etc.). These types of defects are also preferentially torn and leave cones of material around crystal defects. These are clouds of impurities that delay the abrasion of the zone, particularly around screw dislocations.

Techniques involved: ion milling and focused ion beam thinning (FIB).

Structural change: Partial or total change in the crystallographic organization of a sample, caused by ionic effects. It can lead to a change in form, amorphization of the crystal network, a lattice change, crystallization, or re-crystallization.

Techniques involved: ion milling and focused ion beam thinning (FIB).

Microstructural change: Partial or total change of the structural organization of the components of a microstructure that can be caused by ionic effects. It can result in changes in morphology, redistributions of phases, changes in crystal structure, precipitation, chemical gradients, formation of new phases, etc.

Techniques involved: ion milling and focused ion beam thinning (FIB).

2.2.1 Secondary Thermal Damage Induced During Ionic Preparation

Fusion: Change of solid–liquid physical state of a solid sample. It can be caused by an increase in temperature during ionic preparation.

Techniques involved: ion milling and focused ion beam thinning (FIB).

Phase transformation: Total or partial transformation of the crystallographic and/or chemical structure of a sample, which can be caused by thermal effects during ionic preparation.

Techniques involved: ion milling and focused ion beam thinning (FIB).

Loss of chemical elements: Selective loss of a chemical element of the sample, which may be caused by secondary thermal effects during ionic preparation.

Techniques involved: ion milling and focused ion beam thinning (FIB).

Amorphization: Partial or total degradation of the chemical bonds of a crystalline sample, leading to its destruction, caused by thermal effects during the ionic preparation.

Techniques involved: ion milling and focused ion beam thinning (FIB).

Demixing: Separation of partially miscible phases of a mixture or alloy caused by thermal effects, leading to the exceeding of the solubility limit of one or more compounds.

Techniques involved: ion milling and focused ion beam thinning (FIB).

2.3 *Chemical Preparation-Induced Artifacts*

Matter displacement: Transport by creeping of either matter on the sample surface or a particular phase of a multiphase material, caused by chemical action.

Techniques involved: electrolytic polishing, chemical polishing, twin-jet electrolytic thinning, full-bath electrolytic thinning, twin-jet chemical thinning, full-bath chemical thinning, and extractive replica.

Selective dissolution: Chemical or electrolytic polishing or etching can reveal phases if one of them dissolves more slowly than another. Selective dissolution can create an undulating surface if the flow of the dissolution layer or bath agitation are not correctly ensured. It can cause roughness if the bath reveals dense atomic planes and form peaks if precipitates or segregation zones (grain boundaries, dislocations, etc.) are sensitive to the bath.

Techniques involved: electropolishing, chemical polishing, twin-jet electrolytic thinning, full-bath electrolytic thinning, twin-jet chemical thinning, full-bath chemical thinning, and extractive replica.

Composition change: Change in the elemental chemical composition of the sample due to a loss or addition of chemical elements tied to contamination or dissolution during preparation.

Techniques involved: electropolishing, chemical polishing, twin-jet electrolytic thinning, full-bath electrolytic thinning, twin-jet chemical thinning, full-bath chemical thinning, and extractive replica.

Structural change: Partial or total change of the crystallographic and chemical organization of a sample, caused by chemical effects. It can result in a change in form, amorphization of the crystal network, a change in network, or crystallization or re-crystallization.

Techniques involved: electropolishing, chemical polishing, twin-jet electrolytic thinning, full-bath electrolytic thinning, twin-jet chemical thinning, full-bath chemical thinning, and extractive replica.

Microstructural change: Partial or total change of the structural organization of microstructure components, caused by ionic effects. It can result in morphology changes, phase distribution, changes in crystal structure, precipitation, chemical gradients, formation of new phases, etc.

Techniques involved: electropolishing, chemical polishing, twin-jet electrolytic thinning, full-bath electrolytic thinning, twin-jet chemical thinning, full-bath chemical thinning, and extractive replica.

2.3.1 Changes Specific to Biological Materials

Structural change: Alterations of structure at different scales, leading to the observation of a sample that is different from the living material.

- Volume change
- Denaturing of components leading to textural changes
- Transformation of the protein gel into a reticulated structure

Techniques involved: chemical fixation.

- Transformation of membrane phospholipids into unbroken lines

Techniques involved: osmic fixation.

Change in the natural contrast: Change induced by the addition of a heavy metal.

Techniques involved: chemical fixation using osmium and "positive-staining" contrast.

Change in the molecular bonds: Creation of new bonds between macromolecules, leading to interpretation errors of reactive sites after labeling.

Techniques involved: chemical fixation using glutaraldehyde.

Protein cross-linking and changes in their spatial conformation: Connection of macromolecular chains by the creation of bridges or chemical bonds with an exogenous compound, i.e., fixative. This leads to the formation of a network with physical–chemical properties different from the initial material. Cross-linking is an irreversible process that constitutes an obligatory artifact resulting from chemical fixation, but which is indispensable to being able to carry out dehydration. Cross-linking leads to a change or loss of reactivity of the primordial macromolecules. It also causes the formation of bonds between molecules that were originally independent.

Techniques involved: chemical fixation and substitution–infiltration–embedding at room temperature.

Change in chemical composition: Loss of compounds: saturated lipids, simple sugars, ions, salts, loss or addition of chemical elements.

Techniques involved: chemical fixation, substitution–infiltration–embedding at room temperature, and "positive-staining" contrast.

Residues: Deposits of matter extrinsic to the sample, which are chemical residues resulting from preparation. They are superimposed on the thin slice and impede observation. Residues are produced by chemical additions during preliminary preparation or during thinning.

Techniques involved: electropolishing, chemical polishing, continuous support film, holey support film, twin-jet electrolytic thinning, full-bath electrolytic thinning, twin-jet chemical thinning, full-bath chemical thinning, extractive replica, chemical fixation, freeze fracture, "positive-staining" contrast, and immunolabeling.

2.3.2 Secondary Thermal Damage Induced During Chemical Preparation

Microstructural change: Partial or total change of the structural organization of microstructure components that can be caused by thermal effects during chemical preparation. It can result in changes in morphology, phase redistribution, crystal-structure changes, precipitation, chemical gradients, formation of new phases, etc.

Techniques involved: electropolishing, chemical polishing, twin-jet electrolytic thinning, full-bath electrolytic thinning, twin-jet chemical thinning, and full-bath chemical thinning.

Changes in phase distribution: Aggregation of chemical elements caused by thermal diffusion or aggregation of several elements of a microstructure that might lead to the disappearance of the initial phases.

Techniques involved: electropolishing, chemical polishing, twin-jet electrolytic thinning, full-bath electrolytic thinning, twin-jet chemical thinning, and full-bath chemical thinning.

2.4 Physical Preparation-Induced Artifacts

Deformation: Change in shape caused by the formation of ice crystals in a sample that should contain only vitreous ice.

This artifact can generate a change in the conformation and distribution of the matter.

Techniques involved: cryo-fixation and frozen hydrated film

Microstructural changes: Partial or total change in the structural organization of the components of a microstructure, caused by thermal effects (low temperatures). It can result in changes in morphology, phase redistribution, changes in crystal structure, precipitation, chemical gradients, formation of new phases, etc.

Techniques involved: cryo-fixation, substitution–infiltration–embedding in cryogenic mode, cryo-ultramicrotomy, and frozen hydrated film.

Segregation of liquid phases: Migration of a solution in a solvent during a change of physical state that occurs too slow.

Techniques involved: fine particle dispersion Technique, cryo-fixation.

2.4.1 Secondary Thermal Damage Induced During Physical Preparation

Deformation: Change in shape due to heating during physical preparation, leading to the re-crystallization of vitreous ice. This artifact can generate a change in matter conformation and distribution.

Techniques involved: cryo-fixation, substitution–infiltration–embedding in cryogenic mode, cryo-ultramicrotomy, frozen suspension film.

Particle aggregation: Agglomeration of particles caused by the mobility of metallic particles deposited on the sample surface, which is created by a thermal effect or charge effect between ionic particle interaction. This leads to the creation of larger particles through coalescence.

Techniques involved: direct replica, indirect replica, freeze fracture, and decoration shadowing, "positive-staining" contrast.

Fusion: Change in solid–liquid physical state of a solid sample.

It can be caused by an increase in temperature during cryogenic preparation.

Techniques involved: cryo-fixation, substitution–infiltration–embedding in cryogenic mode, and frozen hydrated film.

Phase transformation: Total or partial transformation of the crystallographic and/or chemical structure of a sample, which can be caused by thermal effects during physical deposition.

Techniques involved: direct replica, indirect replica, decoration-shadowing, and "negative-staining" contrast.

Frost: Condensation of atmospheric water vapor on a very cold thin slice.

Techniques involved: cryo-ultramicrotomy and frozen hydrated film.

3 Artifacts Induced During TEM Observation

The investigation of microstructure using electron microscopy involves the interaction of the electron beam with the sample. This interaction irradiates the sample, creates charges in the slice, and can produce heat. The result is physical or chemical changes or the destruction of the material. All of these effects are a function of the acceleration voltage, the dose of electrons received, the size of the zone irradiated, and of course, the nature of the sample. The different types of degradation include artifacts caused by observation and secondary damage caused by thermal effects during observation.

3.1 Artifacts Not Linked to Thermal Damages

Dehydration: The principal effect of the vacuum in the microscope is the instantaneous dehydration of materials in liquid solution or hydrated materials. We cannot work directly on these samples; they must either be immobilized or the water must be extracted.

On samples that are more lightly hydrated and chemically unstable (polymers, biological materials, non-stoichiometric oxides, etc.), the vacuum combined with irradiation can lead to the loss of light elements in a structure. This subsequently leads to phase transformations, resulting in significant changes in shape, which may ultimately result in the complete loss of structure.

Charging: If the charge dissipation is insufficient, the electrons and ions produced under the electron beam cause the appearance of a localized current proportionate to the intensity of the irradiation. This effect results in the instability of the sample under the beam.

The effects are even greater if the beam is focused and the material is insulating, i.e., if it does not allow the flow of electronic charges on the material via the support grid and the specimen holder.

One way to reduce this effect is to deposit a carbon (or metal) film on the surface of the sample.

Destruction: Sample destruction can be caused by the combined effects of irradiation and temperature rise during observation.

Irradiation results in the degradation of the sample, the appearance of stresses causing sample contraction, and the formation of cracks that eventually destroy the sample.

The electron–matter interaction produces heat by plasmon deexcitation. The temperature rise is proportionate to the current density of the beam and is inversely proportionate to the thermal conductivity of the sample.

These effects can result in the amorphization of the material, crystallization, phase demixing, and sample destruction. These effects can be reduced or eliminated by the use of a liquid-nitrogen-cooled specimen holder.

During electronic irradiation, atomic displacement reduces when the incident kinetic energy increases, i.e., when the TEM acceleration voltage increases. Therefore, at high energy, the effective cross section of electron interaction with the matter is greater. The resulting elastic diffusion only leads to weak atomic displacement. For most elements, the probability of atomic displacement is 100 times less than the probability of all elastic interactions. Voltages ranging between 100 and 200 keV represent the best practical compromise for all materials in order to limit irradiation damage.

The contrast of samples composed of light elements is even lower if the acceleration voltage is high. Voltages ranging between 75 and 120 keV represent the best practical compromise for organic materials.

Contamination: Contamination occurs under the action of the electron beam and is caused by the combustion of hydrocarbons present in the TEM column or on the surface of the sample. Hydrocarbons essentially come from oil vapors from the distribution pumps. The hydrocarbon deposits formed are composed of carbon and are more significant when working under a focused probe and an intense probe. Contamination is greatest in a STEM. It results in the formation of dark zones that cover the irradiated portion. Contamination is reduced by an anti-contamination cold trap placed near the sample, in the air gap of the objective lens. This contamination is irreversible and introduces a carbon signal into the chemical analysis. It increases when the TEM vacuum decreases.

In cryomicroscopy, the specimen is very cold and can be contaminated very quickly. In this case, the anti-contamination trap must be placed very close to the specimen.

3.2 Secondary Thermal Damage

Amorphization: Partial or total degradation of the chemical bonds of a crystalline sample, resulting in its destruction. It is caused by thermal effects during the electronic irradiation.

The main effect of irradiation is amorphization of the sample by the breaking of bonds and the loss of elements. It is greater in thin slices prepared by ion milling, which leads to the diffusion of light atoms. In the case of organic materials, electronic irradiation produces changes in the chemical composition by reticulation and the preferential degradation of particular bonds (C–H, C–C, COOH, C–NH_2). The loss of mass with the formation of volatile compounds is often a consequence with polymer materials.

Another effect is annealing by the electron beam, resulting in an atomic surface rearrangement. This chemical disorder of the surface is initiated by diffusion phenomena during ion milling. It is even more significant if the material has a complex phase diagram.

Phase transformation: Total or partial transformation of the crystallographic and/or chemical structure of a sample, which can be caused by thermal effects during electronic irradiation.

Loss of chemical elements: Selective loss of a sample's chemical element, which may be caused by secondary thermal effects during electronic irradiation.

Migration: Collective displacement of atoms that can lead to a migration of phases or interfaces, etc.

Fusion (or sublimation): Change of physical state (solid–liquid) of a solid sample. It can be caused by an increase in temperature during TEM observation.

4 Examples of Artifacts

4.1 Artifacts Induced by the Tripod Polishing Technique

Fig. 6.1 Bright-field TEM image of an MgO single crystal (ceramic material), prepared by the tripod polishing technique, showing cracks due to cleavages in the sample (*J. Ayache, CNRS-UMR8126-IGR, Villejuif*)

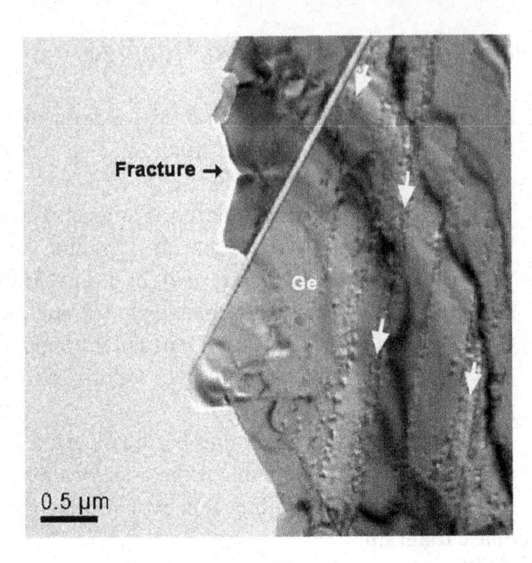

Fig. 6.2 Bright-field TEM image of a germanium cross-section sample (semiconductor) prepared by the tripod polishing technique, showing a tearing of matter due to directional mechanical polishing (*white arrows*) (*C. Dieker, EPFL-CIME, Lausanne*)

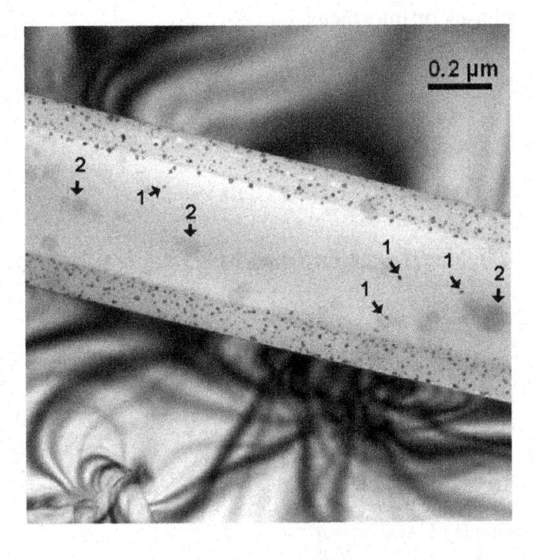

Fig. 6.3 Bright-field TEM image of a cross-section of a SiO_2/Au multilayer on a Si substrate (mixed–composite material) prepared by the tripod polishing technique, showing: (1) displacement of gold particles in the resin; (2) residues of silicon used for final polishing (*S. de Chambrier. A. Schüler, EPFL-LESO, Lausanne*)

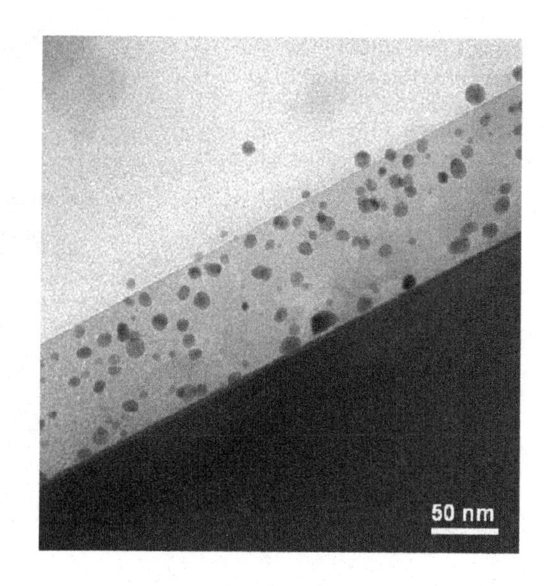

Fig. 6.4 Same sample as in Fig. 6.3, at a higher magnification (*S. de Chambrier, A. Schüler, EPFL-LESO, Lausanne*)

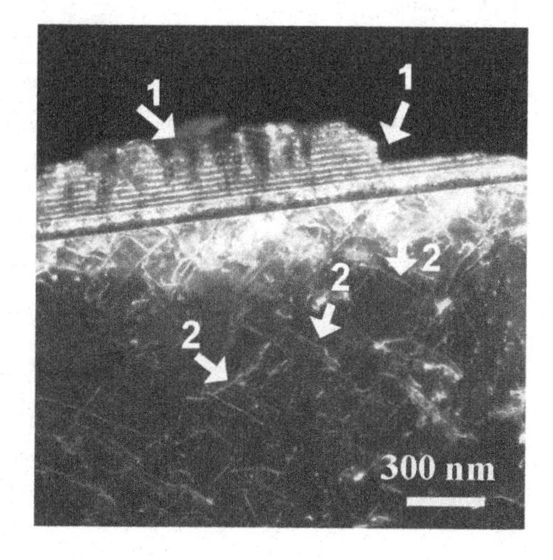

Fig. 6.5 Dark-field TEM image of $YBa_2Cu_3O_7/SrTiO_3$ (composite ceramic cross section) prepared by the tripod polishing technique, showing: (1) tearing of matter in the multilayer and (2) dislocations in the substrate, induced by mechanical polishing (*J. Ayache, CNRS-UMR8126-IGR, Villejuif*)

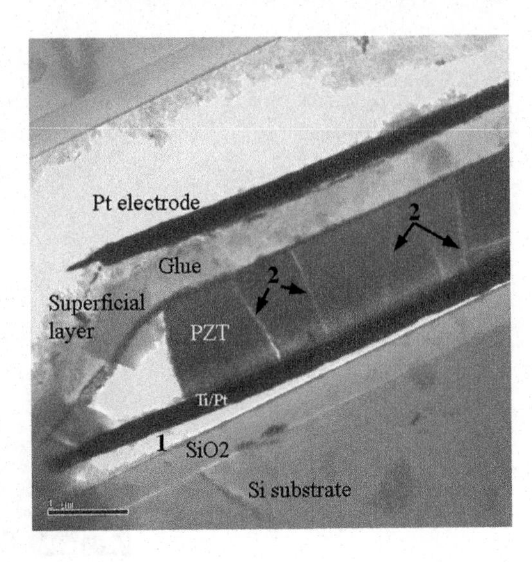

Fig. 6.6 Bright-field TEM image of a cross-section of $Si/SiO_2/Ti/Pt/PZT/Pt$ material prepared by the tripod polishing technique, showing: (1) fracturing of the metal-ceramic $SiO_2/Ti/Pt$ interface of the multilayer material and (2) tearing of matter in the PZT layer, caused by mechanical polishing (*J. Ayache, CNRS-UMR8126-IGR, Villejuif*)

4.2 Artifacts Induced by the Ultramicrotomy Technique

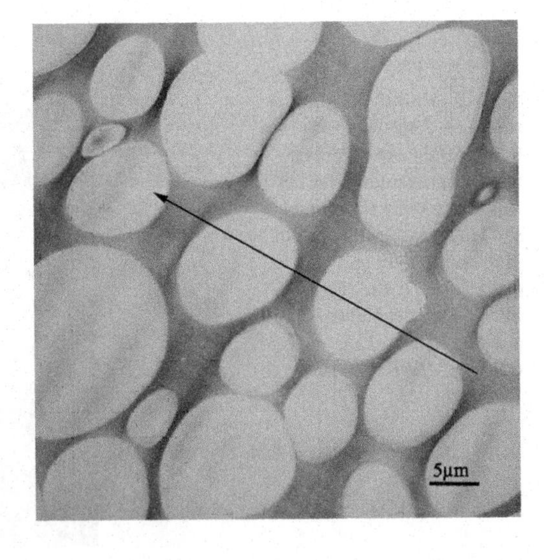

Fig. 6.7 Biphase polymer prepared using ultramicrotomy, showing vibrations and morphology change perpendicular to the sectioning direction (*arrow*) due to compression (*A. Rivoire, EZUS-UCB Lyon 1*)

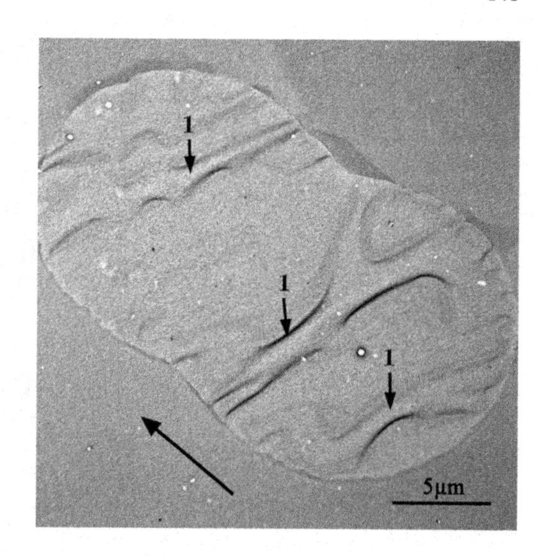

Fig. 6.8 Bright-field TEM image of another biphase polymer (polymer–composite) section prepared using ultramicrotomy, showing folds on a nodule in the matrix, which is of a different hardness. The folds (1) are found on the softer polymer (*A. Rivoire, Ezus-UCB Lyon 1*)

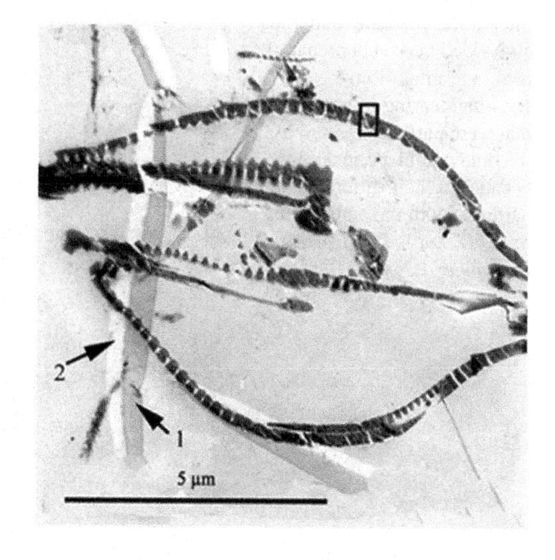

Fig. 6.9 Bright-field TEM image of a diatom section prepared using ultramicrotomy after embedding in resin, showing (1) folds and (2) tears of the embedding resin. This is an alga in which only the mineralized shell has been preserved. This shell is ornamented with fine ribs presented in the form of regularly spaced thicknesses of SiO_2 (*small black* masses) separated by pores (*D. Badaut, EOST Strasbourg*)

Fig. 6.10 Higher magnification of one area of the previous image, showing fractures and detachments in a side of the siliceous wall of the diatom. Silica is a hard and brittle material that is difficult to section (*D. Badaut, EOST Strasbourg*)

Fig. 6.11 Bright-field TEM image of a cross-section of SiO_2 fibers (ceramic material) embedded in a resin prepared using ultramicrotomy, showing: tearing of two fibers that were partially ejected by the diamond blade in cutting because of their difference in hardness with the embedding resin (*D. Laub, EPFL-CIME, Lausanne*)

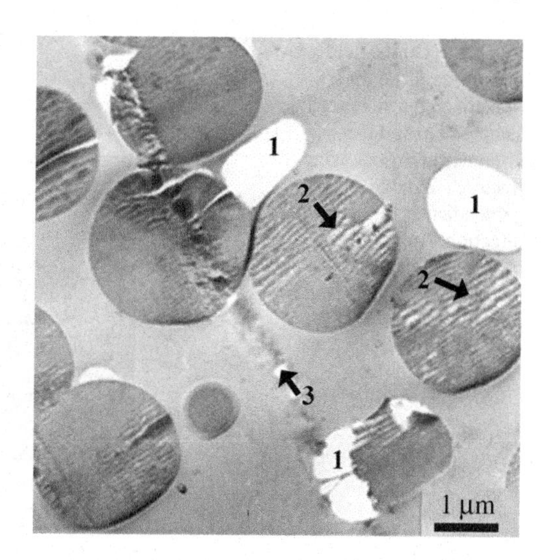

Fig. 6.12 Bright-field TEM image of isolated carbon particles (semi-crystalline) embedded in a resin prepared using ultramicrotomy, showing: (1) tearing due to the difference in hardness of the spheres with the embedding resin, (2) microfissures in the spheres, caused the compression due to the blade followed by a rupture of the material, (3) a large knife ridge perpendicular to the cracks (*J. Ayache, CNRS-UMR8126-IGR, Villejuif*)

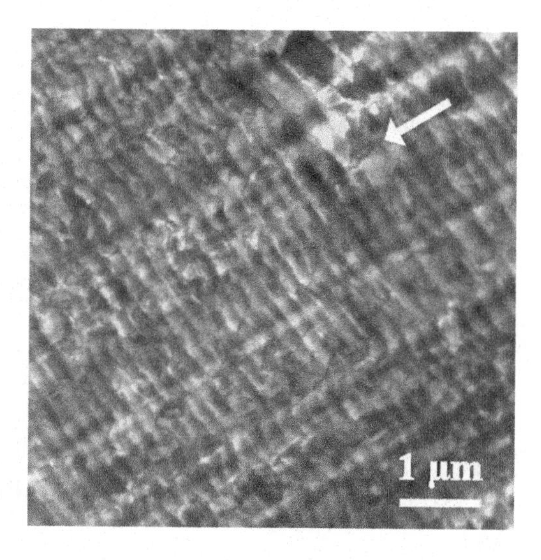

Fig. 6.13 Bright-field TEM image of a tin (*metal*) particle, embedded and prepared by ultramicrotomy, showing: (1) parallel knife ridges in the cutting direction (*arrow*), (2) compression and strain hardening of the material highlighted by the micro-undulations perpendicular to the sectioning direction (*arrow*) (*D. Laub, EPFL – CIME Lausanne*)

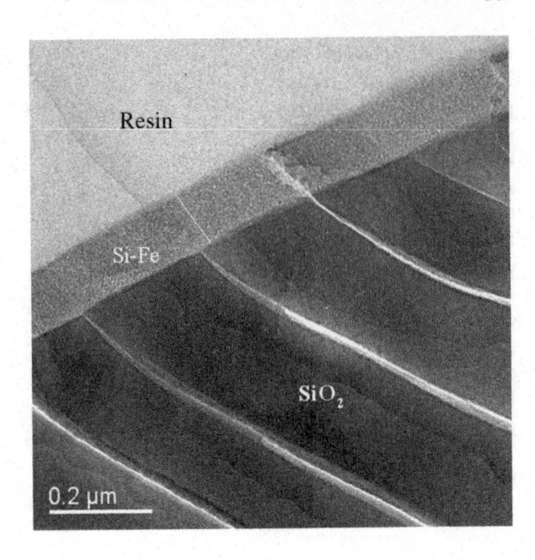

Fig. 6.14 Bright-field TEM image of a Si-Fe/SiO_2 (mixed–composite) particle sample prepared using ultramicrotomy, showing fractures in the SiO_2 perpendicular to the cutting direction. However, observation of the Fe–Si and the Fe–Si/SiO_2 interface is possible (*D. Laub, EPFL – CIME Lausanne*)

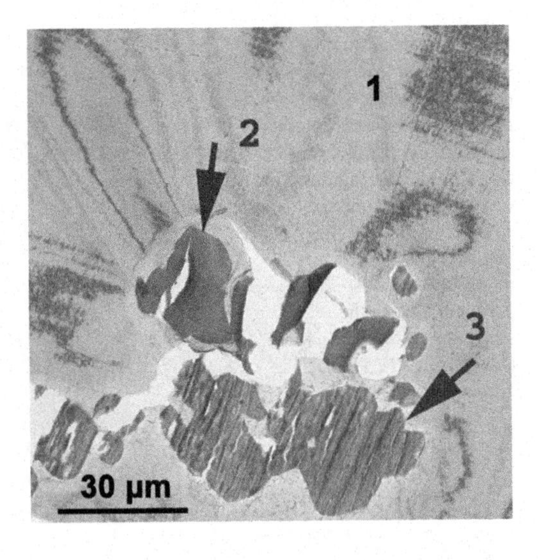

Fig. 6.15 Bright-field TEM image of an aluminum-based alloy (metallic material) prepared without embedding on the bulk material using ultramicrotomy, showing tearing of matter due to the heterogeneity in hardness of phases (2) and (3) with the strain-hardened matrix (1). More significant strain hardening can be seen in phase (3) (*J. Ayache, CNRS-UMR8126-IGR, Villejuif*)

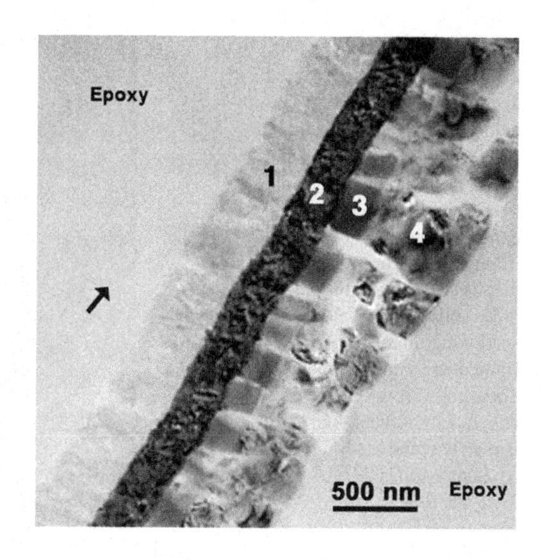

Fig. 6.16 Bright-field TEM image of a cross-section of a multilayer DLC/Ti/Si_3N_4/Si (mixed–composite) material embedded in an epoxy resin and prepared by ultramicrotomy, principally showing fractures in the Si_3N_4 layer (3) and Si (4) perpendicular to the cutting direction (*arrow*) and undulations of compression visible in the DLC (1) and Ti (2) (*H.Gnaegi, Diatome, Bienne*)

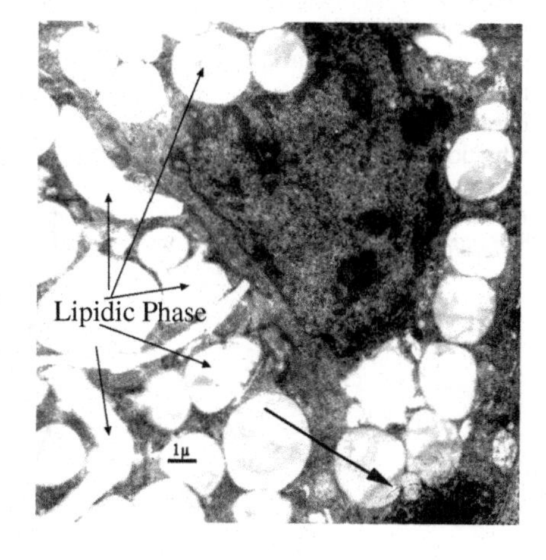

Fig. 6.17 Bright-field TEM image of a cell prepared using chemical fixation, embedding in epoxy resin, and ultramicrotomy, showing cutting residues (*large arrow*) which are superimposed over the structures to be observed. Saturated lipids form white spheres (*small arrows*) (J. *Boumendil CMEABG, UCB Lyon 1*)

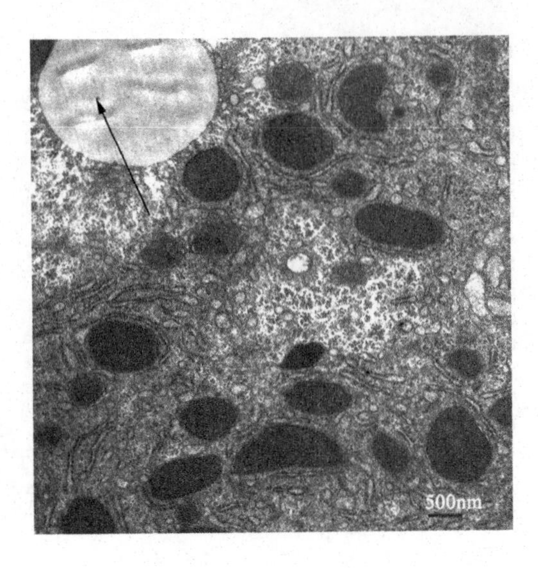

Fig. 6.18 Bright-field TEM image of a liver cell prepared using chemical fixation, embedding in epoxy resin, and ultramicrotomy, showing a triglyceride phase (*arrow*) that has undergone compression phenomena: ridges and matter displacement, resulting in holes followed by folds in the sectioning direction (indicated by the *arrow direction*). From this we conclude that the softer phase is poorly embedded (*J. Boumendil, CMEABG UCB Lyon 1*)

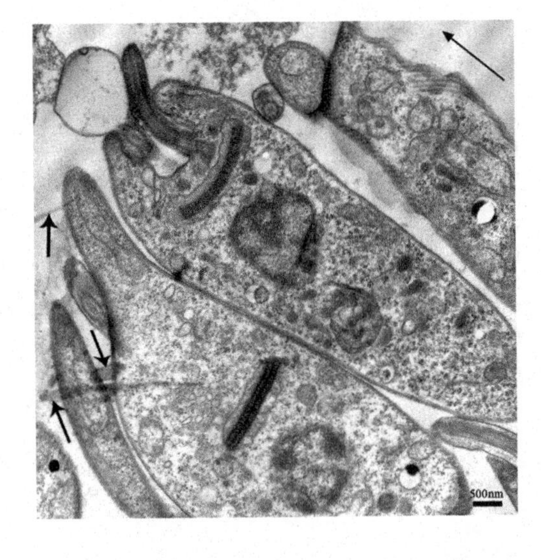

Fig. 6.19 Bright-field TEM image of a slice of a protozoan embedded in epoxy resin and prepared by ultramicrotomy showing vibrations (*arrows*) and folds (*arrow heads*) (*E. Delain, CNRS-UMR8126-IGR, Villejuif*)

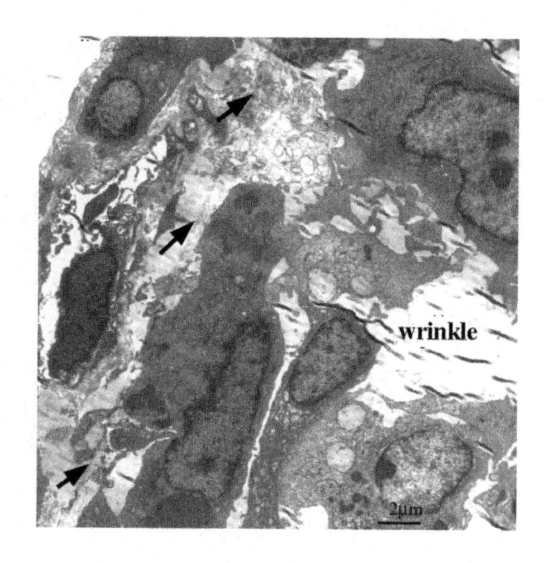

Fig. 6.20 Bright-field TEM image of a slice of placenta prepared by ultramicrotomy showing folds perpendicular to the cut and a long ridge (marked by *arrows*) in the cutting direction (*J. Boumendil, CMEABG UCB Lyon 1*)

4.3 Artifacts Induced by the Freeze-Fracture Technique

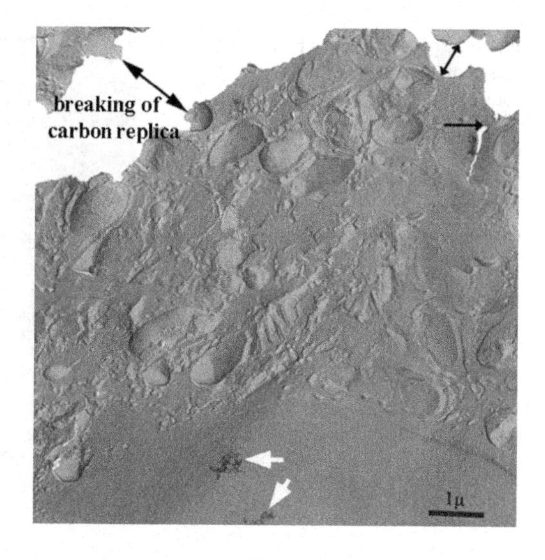

Fig. 6.21 Bright-field TEM image of a liver cell replica, prepared using the freeze-fracture technique, showing a rupture of the replica produced when the replica was washed and the presence of carbon residues (*white arrow heads*) (*A. Rivoire, EZUS-UCB Lyon 1*)

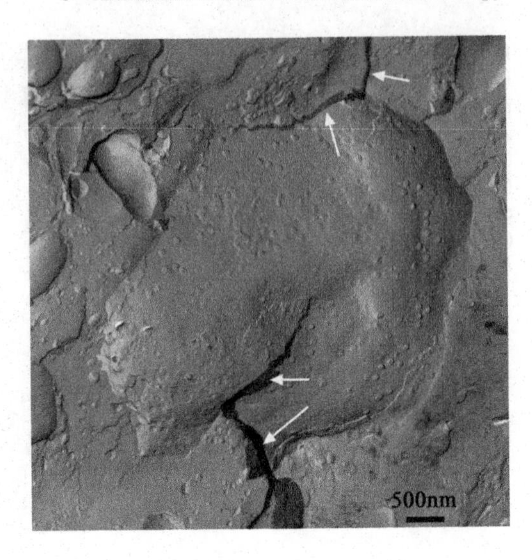

Fig. 6.22 Bright-field TEM image of a replica of a cell showing the nucleus prepared using the freeze-fracture technique. Folds of the replica (*white arrows*) are evident in the areas of greatest relief of the sample *(A. Rivoire, EZUS-UCB Lyon 1)*

4.4 Artifacts Induced by Ion Milling or FIB

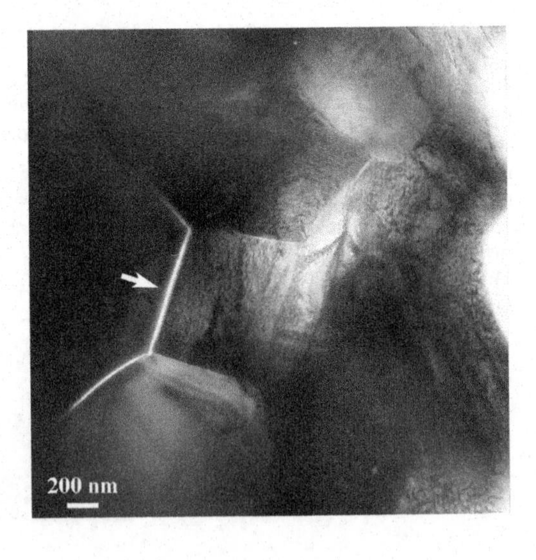

Fig. 6.23 Bright-field TEM image of a polycrystalline ceramic, $YBa_2Cu_3O_7$, prepared using the ion milling technique, showing an intergranular crack (*arrow*) (*J. Ayache, CNRS-UMR8126-IGR, Villejuif*)

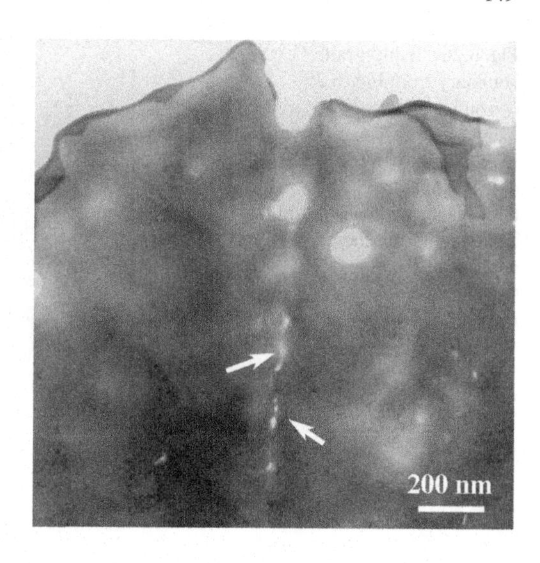

Fig. 6.24 Bright-field TEM image of an MgO bicrystal grain boundary (ceramic), prepared by the ion milling technique, showing a preferential thinning at the grain boundary (*arrows*) (*J. Ayache, CNRS-UMR8126-IGR Villejuif*)

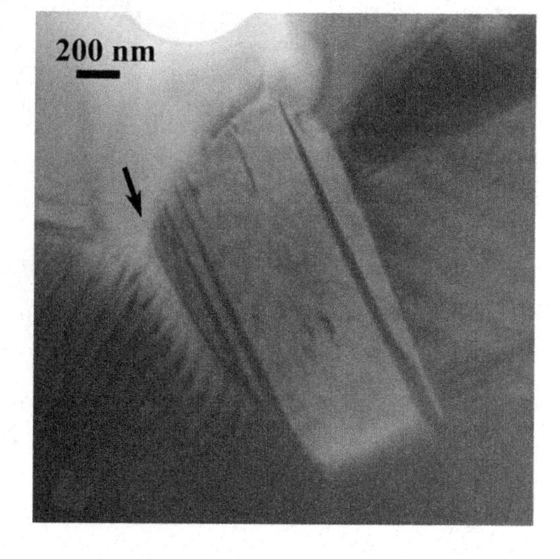

Fig. 6.25 Bright-field TEM image of a polycrystalline ceramic, $YBa_2Cu_3O_7$, prepared using ion milling, showing differential ionic abrasion of the grains due to differences in crystallographic orientation (*J. Ayache, CNRS-UMR8126-IGR, Villejuif*)

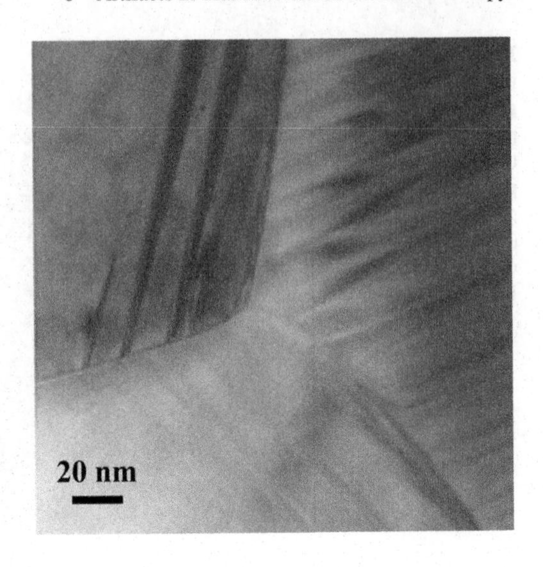

Fig. 6.26 Triple-grain boundary from Fig. 6.25 (*arrow*) at a higher magnification. The difference in thickness between grains does not enable high-resolution observation and quantitative chemical analysis of the interfaces (*J. Ayache, CNRS-UMR8126-IGR, Villejuif*)

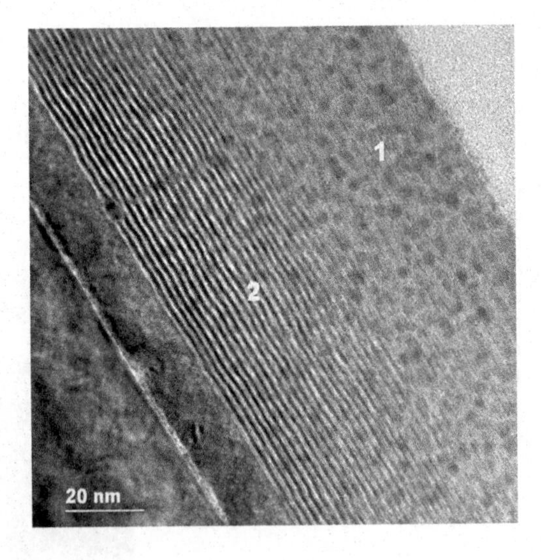

Fig. 6.27 HRTEM image in cross-section of a multilayer Ru/Zr (2) on a $SrTiO_3$ substrate, prepared using the tripod polishing method followed by final ion milling. The external Ru/Zr layers (1) were destroyed by ion milling (*D. Laub, EPFL-CIME, Lausanne*)

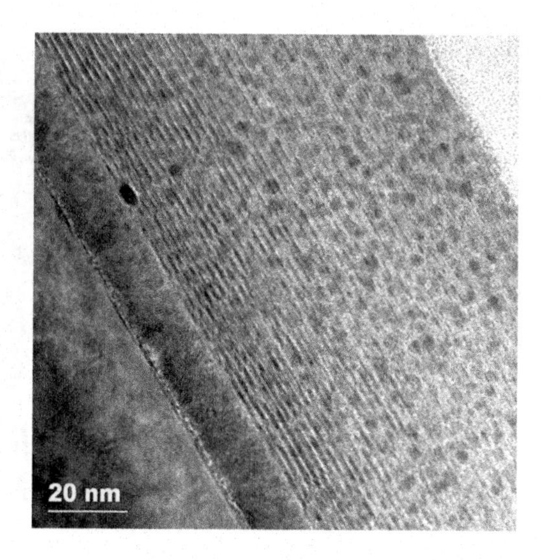

Fig. 6.28 Bright-field TEM image. This is the same sample as shown in Fig. 6.27, but instead focuses on another area where all of the layers were damaged and covered with redeposited particles. Dark particles on the layers have been abraded and redeposited on the material during ion milling. They make observation difficult (*D. Laub, EPFL-CIME, Lausanne*)

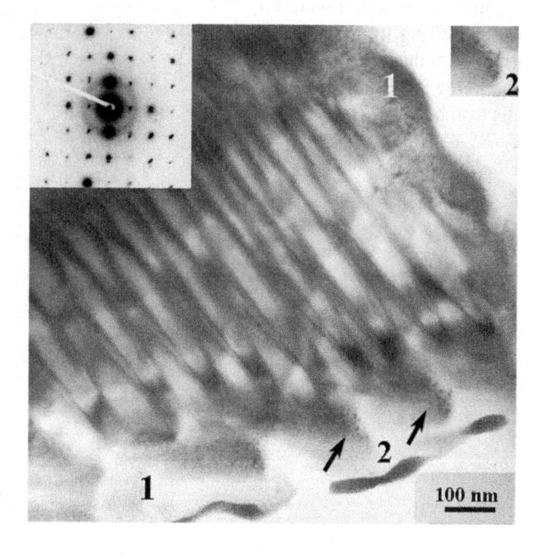

Fig. 6.29 Bright-field TEM image of a $YBa_2Cu_3O_7$ ceramic, prepared using the ion milling technique, showing: (1) amorphization of the material at the edges of the twin walls in the very thin zones, (2) precipitates (*arrows*) due to the demixing of $YBa_2Cu_3O_7$, first under the action of the ion beam and then the electron beam (*J. Ayache, CNRS-UMR8126-IGR, Villejuif*)

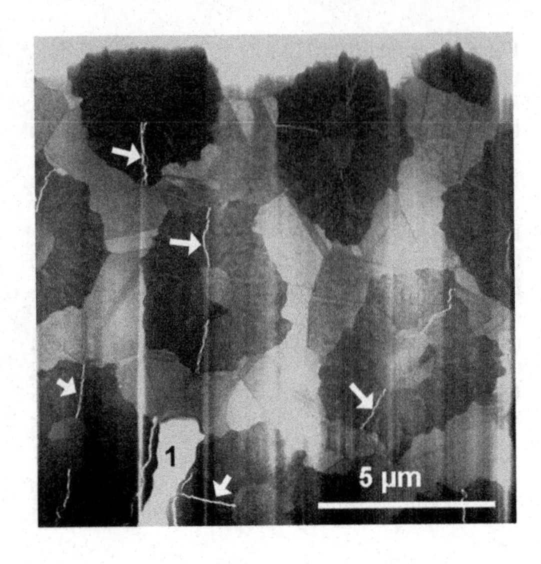

Fig. 6.30 Bright-field TEM image of a multifilament Nb_3Sn sample in a bronze matrix prepared using the FIB method, showing a hole in the slice due to differential thickness ("curtain effect") (1) and fractures in the filaments due to stresses to the material during preparation (*arrows*) (*M. Cantoni, EPFL-CIME, Lausanne*)

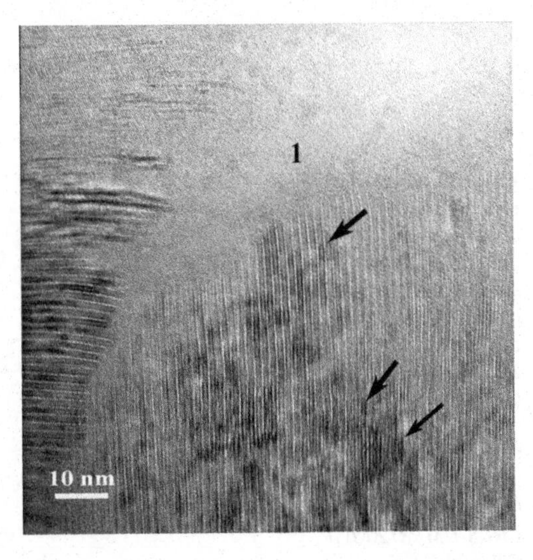

Fig. 6.31 HRTEM image of $YBa_2Cu_3O_7$ ceramic prepared using the ion milling technique, showing amorphization of the material (1) in the very thin areas, and irradiation damage (*arrows*) due to the ion beam and electron beam (*J. Ayache, CNRS-UMR8126-IGR, Villejuif*)

Fig. 6.32 HRTEM image of the polycrystalline $YBa_2Cu_3O_7$ prepared using the ion milling technique, showing (1) a selective thinning at a grain boundary, where the grains present a strong disorientation and contain an intergranular phase, and (2) damage from irradiation under the electron beam (*J. Ayache, CNRS-UMR8126-IGR, Villejuif*)

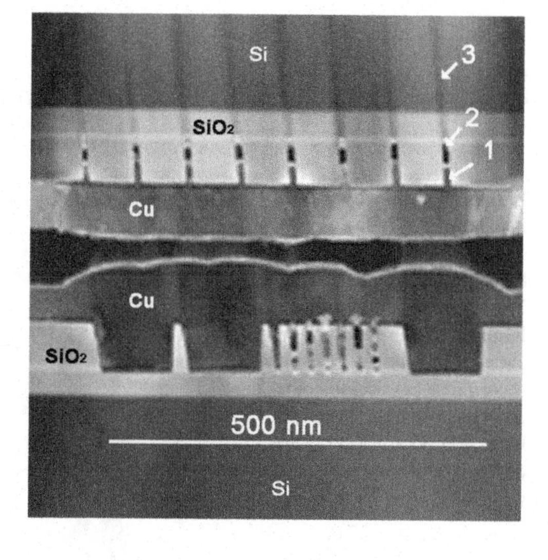

Fig. 6.33 Bright-field TEM image of a $Cu/SiO_2/Si$ sample prepared cross-sectionally using the tripod polishing technique, followed by ion bombardment in single sectorial rotation. Copper (1) is not found in the bottom of the gaps (2). The dark lines behind the gaps (3) are thicker regions, resulting from the abrasion redeposition of Cu (*D. Laub, F. Cosandey, Rutgers University, Piscataway, NJ, USA*)

4.5 Artifacts Induced by the Substitution–Infiltration–Embedding Technique

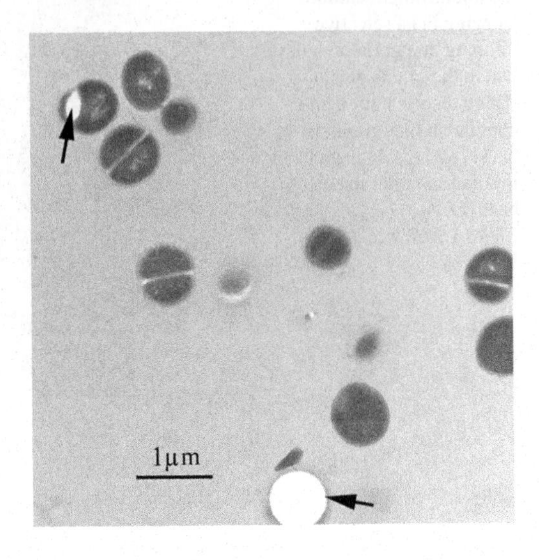

Fig. 6.34 Bright-field TEM image of a slice of *Staphylococcus aureus* embedded first in agar and then in epoxy resin. This image shows holes in the resin due to incomplete dehydration and embedding (*arrows*) (*J. Boumendil, CMEABG UCB Lyon 1*)

4.6 Artifacts Induced by Chemical Fixation

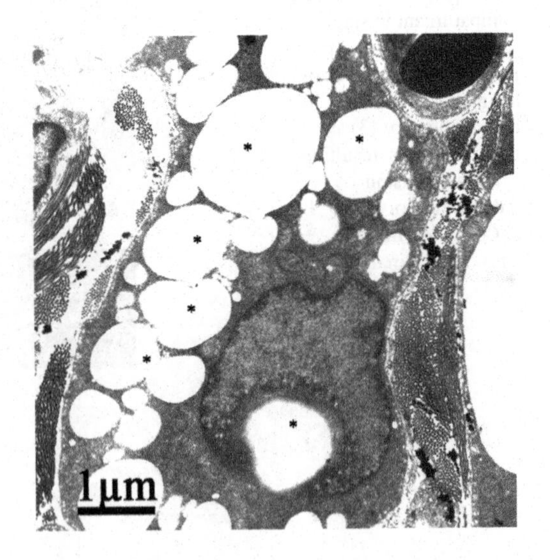

Fig. 6.35 Bright-field TEM image of an adipocyte cell, showing a loss of lipids (*white holes* with *asterisks*) located in the cytoplasm. The lipids disappeared during chemical fixation. As sectioning was possible, the resin has filled the residual holes, but does not show any contrast (*J. Boumendil, CMEABG UCB Lyon 1*)

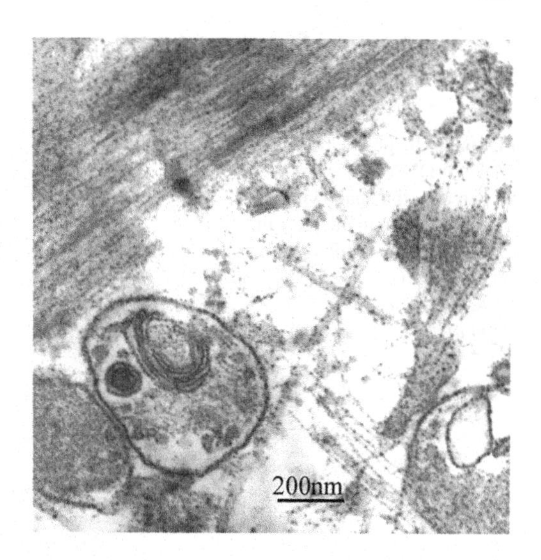

Fig. 6.36 Bright-field TEM image showing a portion of a cell that has undergone a double chemical fixation, embedding in epoxy resin and positive staining. The small dark grains visible everywhere are precipitates of metallic osmium, which are not bound to the structures and constitute an artifact (*J. Boumendil, CMEABG, UCB Lyon 1*)

4.7 Artifacts Induced by the Extractive-Replica Technique

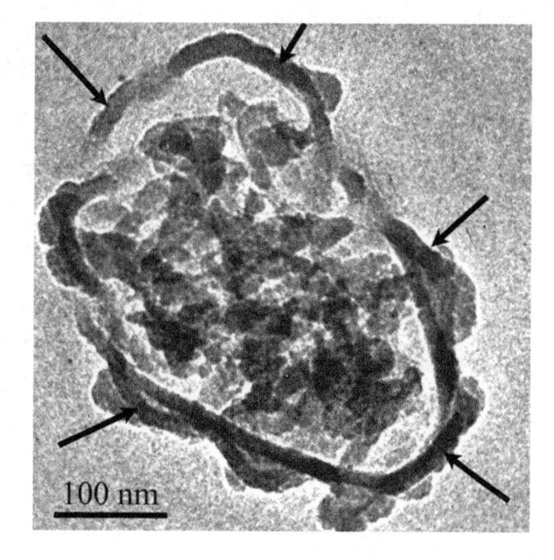

Fig. 6.37 Bright-field TEM image of a catalyst containing 0.2% particles of platinum on an alumina substrate (Pt/Al_2O_3), prepared using the extractive-replica technique. The final chemical attack with fluorhydric acid was not sufficient and, consequently, the interior of the alumina substrate was not dissolved. The outside edges of the alumina substrate are indicated by the *arrows* (*G. Ehret, IPCMS, Strasbourg*)

4.8 Artifacts Induced by the Shadowing Technique

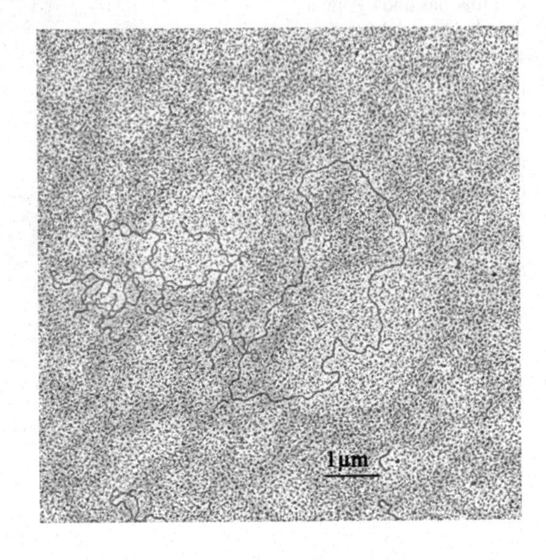

Fig. 6.38 Bright-field TEM image of DNA molecules prepared by the shadowing technique, showing defects in the homogeneity of the support film (*E. Delain, CNRS-UMR8126-IGR, Villejuif*)

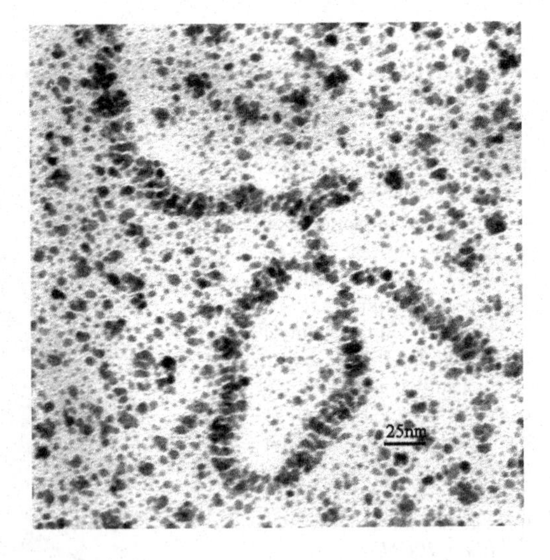

Fig. 6.39 Bright-field TEM image of a fragment of DNA molecules prepared by the shadowing technique, showing an aggregation of the platinum shadowing particles (*E. Delain, CNRS-UMR8126-IGR, Villejuif*)

4.9 Artifacts Induced by the "Positive-Staining" Contrast Technique

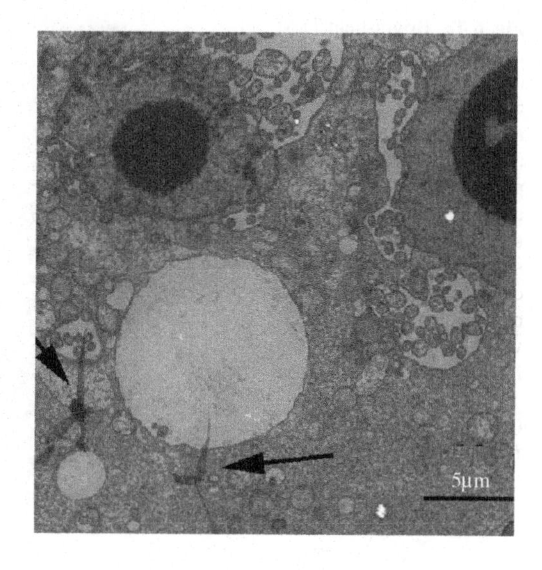

Fig. 6.40 Bright-field TEM image of a slice of embryonic cells from a seedling showing needle-shaped uranyl precipitates (*J. Boumendil CMEABG, UCB Lyon 1*)

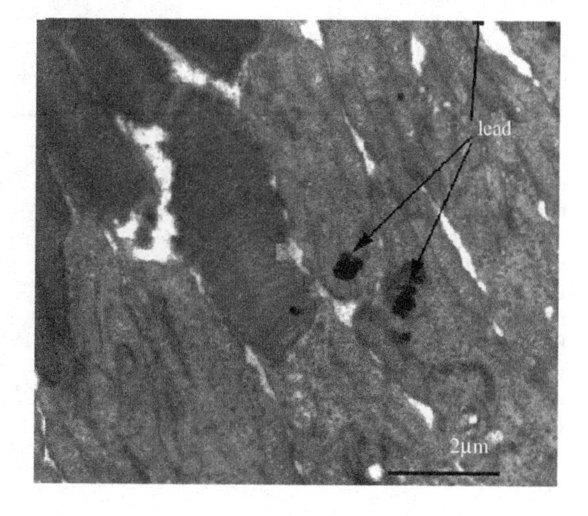

Fig. 6.41 Bright-field TEM image of a slice of a cell showing spherical lead precipitates (*J. Boumendil CMEABG, UCB Lyon 1*)

4.10 Artifacts Induced by the Cryofixation Technique

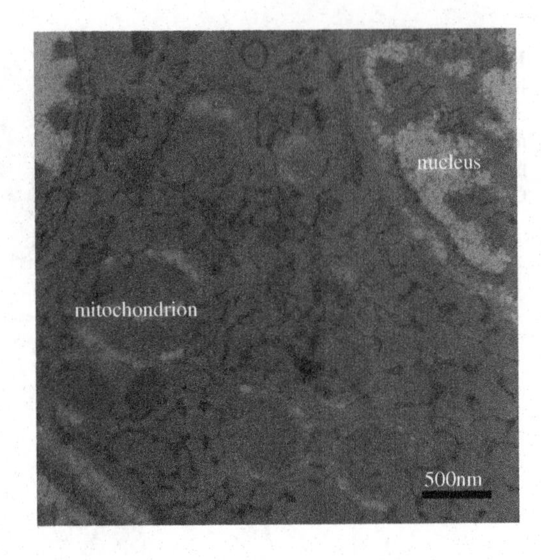

Fig. 6.42 Bright-field TEM image of a rat liver cell slice cryofixed using a projection on cooled copper at 20 μm around the freezing front, then freeze substitution in the presence of osmium and embedding at low temperatures in resin. It shows microstructural changes due to the presence of ice crystals throughout, which disturb the interpretation of the image. The freezing was too slow (*J. Boumendil, CMEABG – UCB Lyon 1*)

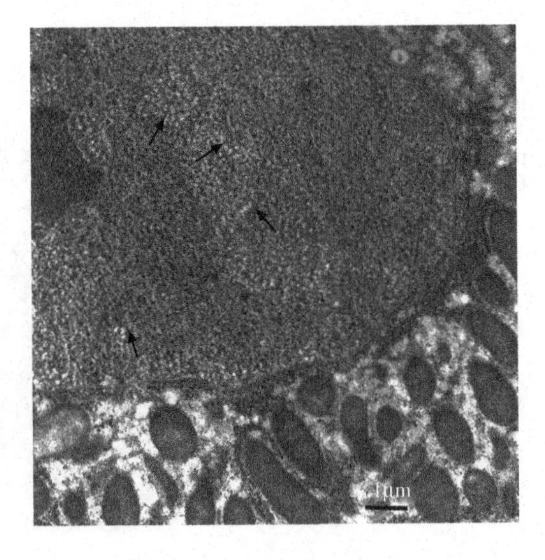

Fig. 6.43 Bright-field TEM image of a rat liver cell slice cryofixed using a projection on cooled copper and at about 10 μm around the freezing front, then freeze substitution in the presence of osmium and embedding at low temperatures in Lowicryl resin. Note the presence of very small crystals of cubic ice (*black arrows*) in the nucleus. The microstructure has been modified; however, the image can still be interpreted (*J. Boumendil, CMEABG – UCB Lyon 1*)

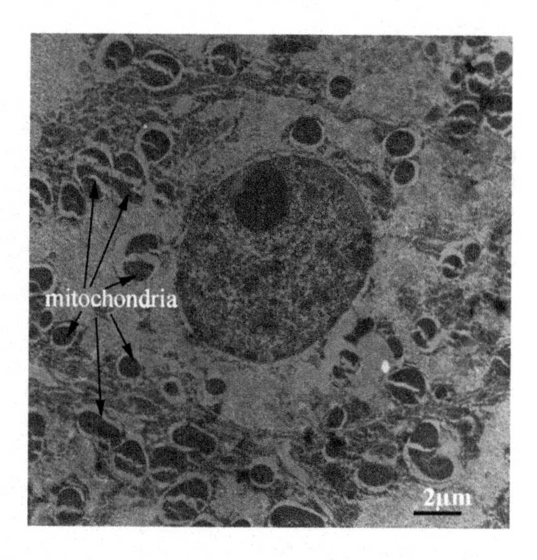

Fig. 6.44 Bright-field TEM image of a slice of liver frozen under high pressure, cryo-dehydrated without OsO_4, cryo-embedded and contrasted using uranyl acetate/lead. This image shows fractured mitochondria, either from the high pressure or due to mechanical tearing when recovering the sample after freezing (*J. Boumendil, CMEABG – UCB Lyon 1*)

4.11 Artifacts Induced by the Fine Particle Dispersion Technique

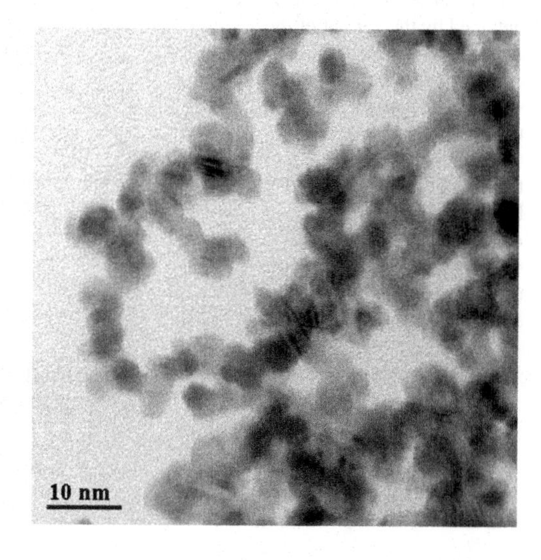

Fig. 6.45 Bright-field TEM image of Pt/Ru/Sn (metal) powder dispersed on a carbon film grid. The particles were dissolved in ethanol, then agitated under ultrasound waves before a drop of supernatent was deposited on the support grid. The particles remain agglomerated (*D. Laub, EPFL-CIME, Lausanne*)

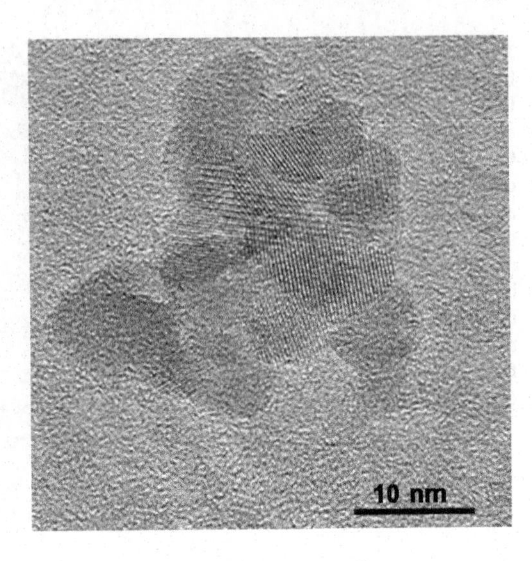

Fig. 6.46 Enlargement of the TEM image from Fig. 6.45. The agglomeration of the particles makes it impossible to precisely measure particle size (*D. Laub, EPFL-CIME, Lausanne*)

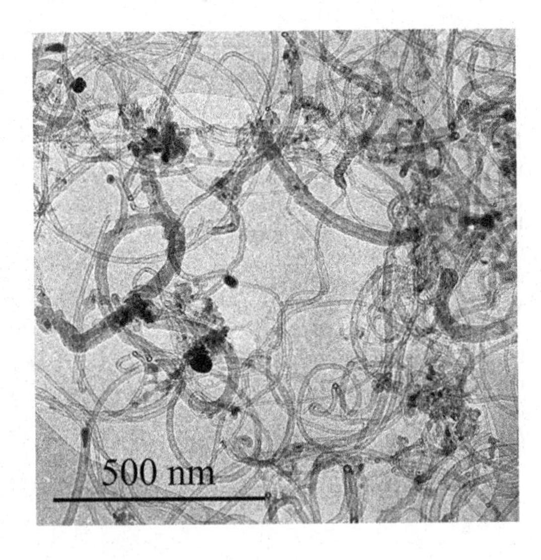

Fig. 6.47 Bright-field TEM image of a dispersion of carbon nanotubes (polymer) obtained by the variant of the crushing technique (i.e., dry crushing between two glass slides). The nanotubes are then transferred to a holey-carbon grid. The tangling of the nanotubes, because of their flexibility, makes their observation difficult at low magnifications (*C.S. Cojocaru, IPCMS, Strasbourg*)

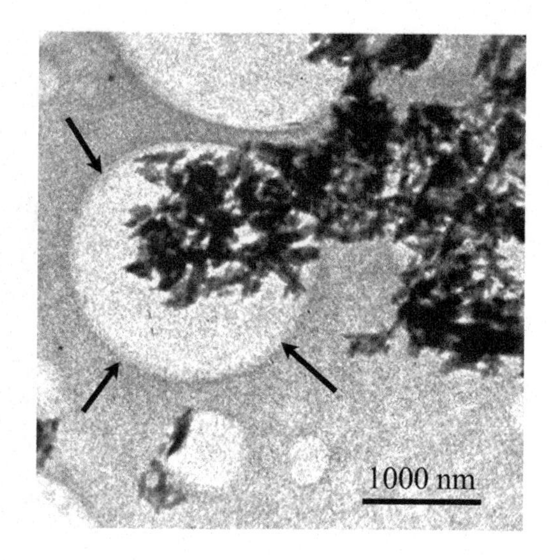

Fig. 6.48 Bright-field TEM image of ZnO (ceramic) particles obtained by dispersion in water and deposited on a holey-carbon grid; the particles are agglomerated in a hole, indicated by the *arrows*, making observation difficult (*K. Lattaud, IPCMS, Strasbourg*)

4.12 Artifacts Induced by the Frozen-Hydrated-Film Technique

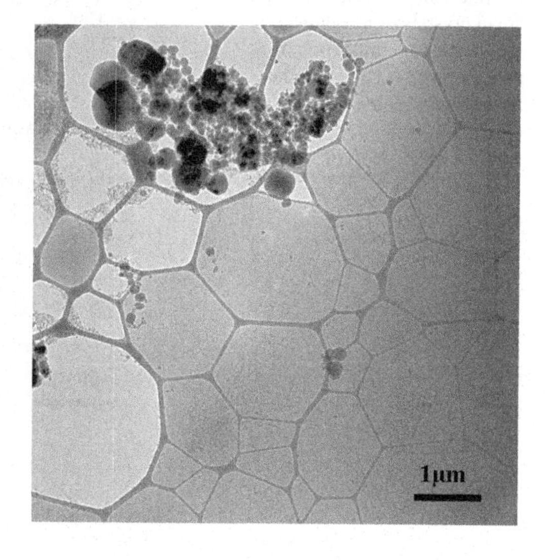

Fig. 6.49 Bright-field TEM image of a frozen hydrated film showing crystals of cubic and hexagonal ice on the carbon holes without a frozen hydrated film (*G. Pehau-Arnaudet, Institut Pasteur-CNRS, Paris*)

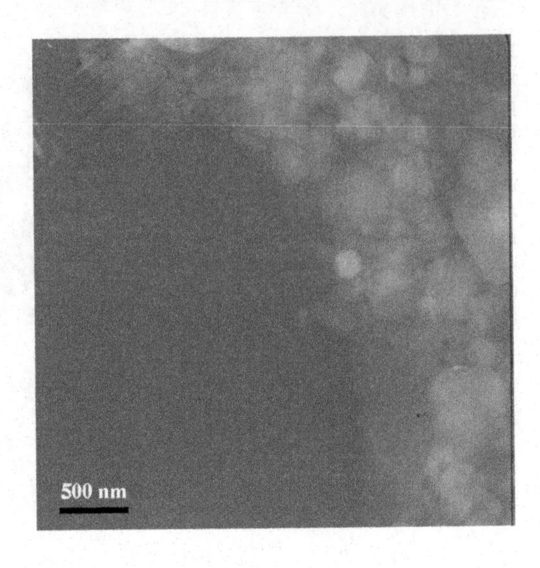

Fig. 6.50 Bright-field TEM image of a frozen hydrated film showing clusters of agglomerated crystals on the edges of a hole on a holey carbon film support. This is due to poor freezing (*S. Baconnais, CNRS-UMR8126-IGR, Villejuif*)

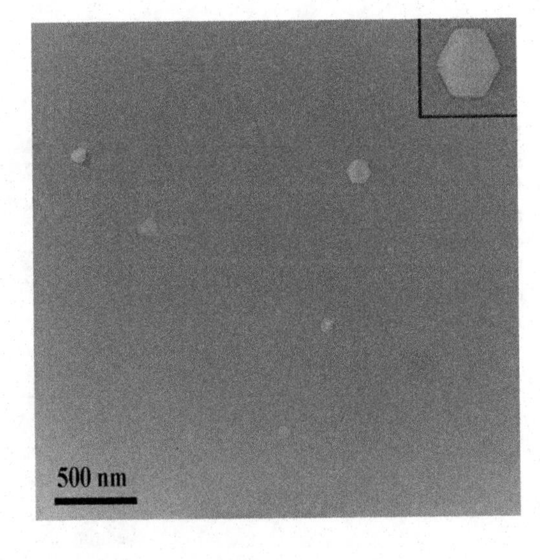

Fig. 6.51 Bright-field TEM image of a continuous frozen hydrated film presenting local small crystals of hexagonal ice. Higher magnification of an ice crystal is shown in the *upper right-hand corner* (*S. Baconnais, CNRS-UMR8126-IGR, Villejuif*)

Fig. 6.52 Bright-field TEM image of vitrified liposomes in a frozen hydrated film, showing a phase-contrast inversion between the ice and the liposome which is due to the absence of osmium or heavy metal (Ur or Pb) commonly used for biological preparation (*G. Pehau-Arnaudet, Institut Pasteur-CNRS, Paris*)

4.13 Artifacts Induced by the "Negative-Staining" Contrast Technique

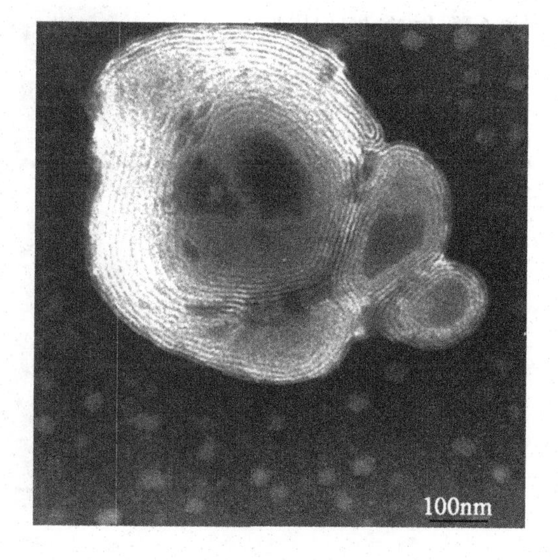

Fig. 6.53 Bright-field TEM image of a liposome suspension under negative staining using uranyl acetate. This image shows residues of the dispersion agent (*white spots*), which can be observed around the liposome and are sometimes superimposed over the liposome structure. They come from a component of the suspension liquid that was not eliminated (*J. Boumendil, CMEABG – UCB Lyon 1*)

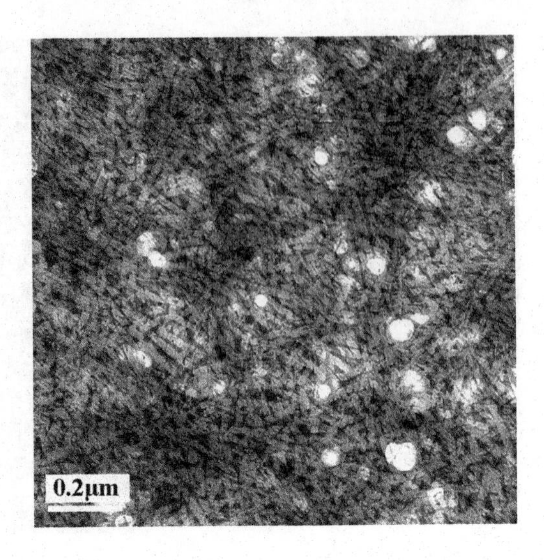

Fig. 6.54 Bright-field TEM image of a suspension of the tobacco mosaic virus (TMV) in 5% uranyl acetate, showing a strong concentration of TMV, which prevents the observation of isolated particles (*P.Y. Sizaret, Université Tours*)

4.14 Artifacts Induced by the Electron Beam

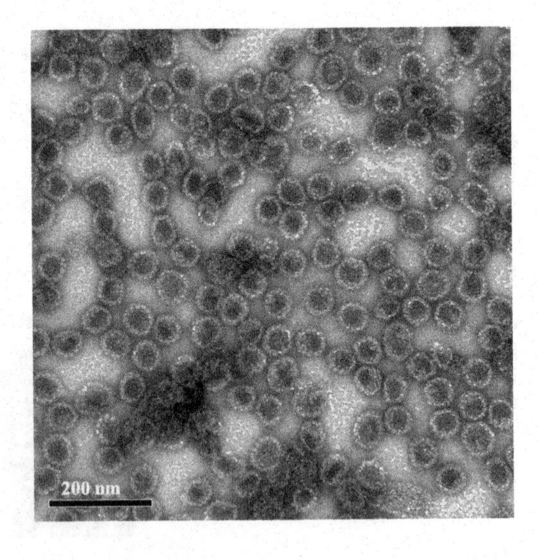

Fig. 6.55 Bright-field TEM image of negative staining of viral pseudo-particles using uranyl acetate, showing the presence of irradiation damage formed under the electron beam, resulting in sublimation of the staining agent that becomes small white dots (*P.Y. Sizaret, Université Tours*)

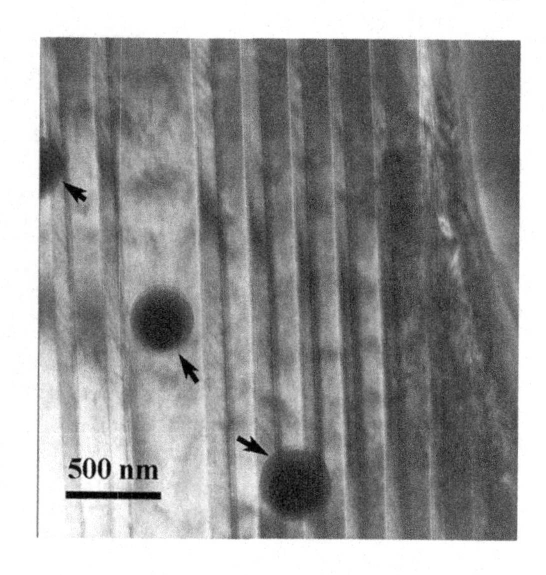

Fig. 6.56 Bright-field TEM image of a superconducting ceramic, $YBa_2Cu_3O_7$, after EDS chemical analysis on the walls of the twins. We can see the traces of carbon contamination (*arrows*) due to the electron beam (*J. Ayache, CNRS-UMR8126-IGR, Villejuif*)

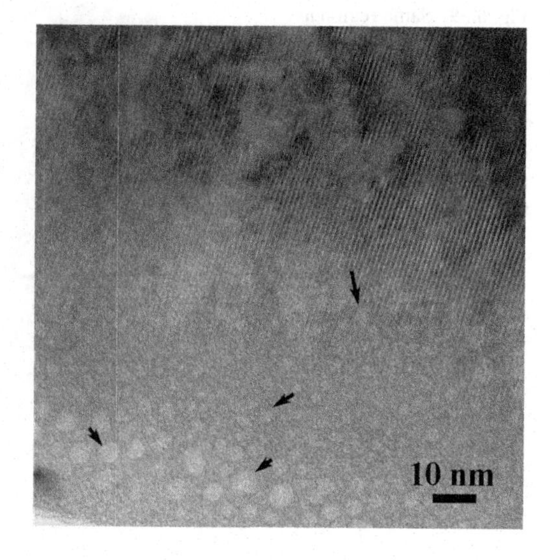

Fig. 6.57 Bright-field TEM image of a superconducting ceramic, $YBa_2Cu_3O_7$, showing irradiation damage under the electron beam. They appear in the form of pores (*arrows*). This damage is encouraged by the strong tendency toward decomposition, due to the non-stoichiometry of these perovskite oxides (*J. Ayache, CNRS-UMR8126-IGR, Villejuif*)

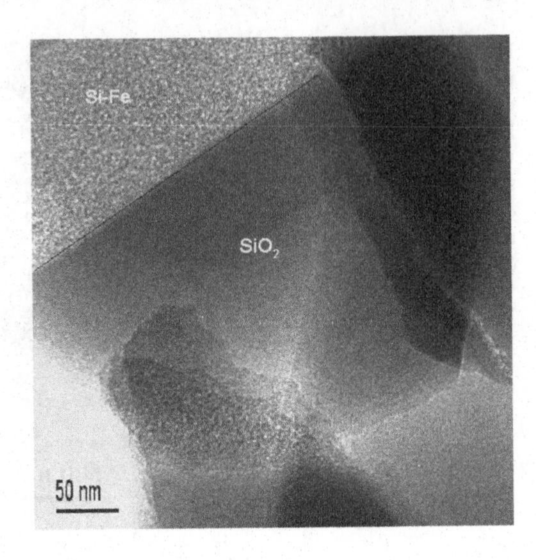

Fig. 6.58 Bright-field TEM image of an interface of Si-Fe/SiO_2 prepared using ultramicrotomy, showing the structure of the SiO_2 after a short exposure to the electron beam (*D. Laub, EPFL-CIME, Lausanne*)

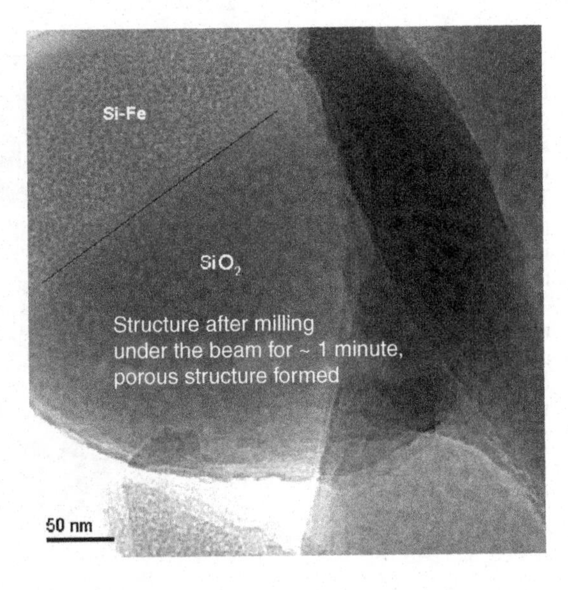

Fig. 6.59 Same region as shown in Fig. 6.58. This image shows the structure of SiO_2 after an approximately 1-min exposure under the electron beam. A porous structure has formed (*D. Laub, EPFL-CIME, Lausanne*)

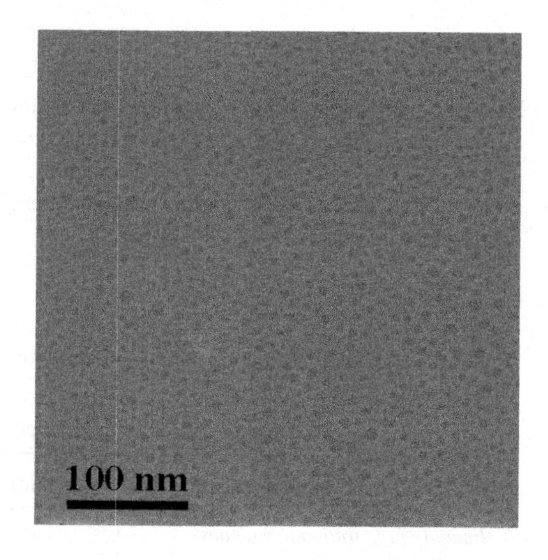

Fig. 6.60 Bright-field TEM image of a frozen hydrated film, showing bubbles from ice melting under the electron beam (*S. Baconnais, CNRS-UMR8126-IGR, Villejuif*)

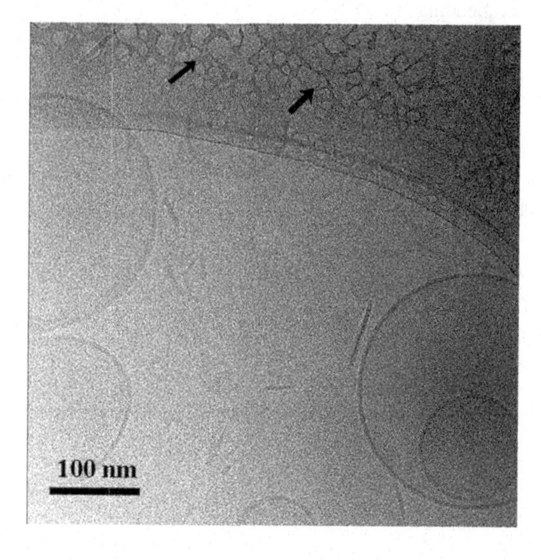

Fig. 6.61 Bright-field TEM image of vitrified liposomes in a frozen hydrated film, showing the effects of irradiation on the ice film (*arrows*) (*G. Pehau-Arnaudet, Institut Pasteur, Paris*)

5 Summary Tables

Two summary tables are presented below (Tables 6.2 and 6.3). These tables list the artifacts caused by preliminary preparation techniques, chemical- and ionic-thinning techniques, mechanical-preparation techniques, replica techniques, techniques specific to divided materials, contrast enhancement techniques, and immunolabeling techniques.

Table 6.2 Table summarizing the various artifacts formed by the preliminary preparation techniques

	Artifacts induced by preparation				
Technique	Mechanical	Ionic	Chemical	Physical	Thermal damage due to preparation
Preliminary preparation techniques					
Sawing	+				
Ultrasonic cutting					
Mechanical polishing	+				+
Dimpling	+				
Electropolishing			+		
Chemical polishing			+		
Sandwich technique					
Embedding			+		+
Substitution–infiltration–embedding at room temperature			+		+
Substitution–infiltration–embedding in cryogenic mode			+	+	
Chemical fixation			+		
Physical fixation: Cryo-fixation				+	
Continuous support films			+	+	
Holey support films			+	+	

Table 6.3 Table summarizing the various artifacts formed by the techniques

Preparation Techniques					
	Type of Artifact				
Technique	Mechanical	Ionic	Chemical	Physical	Thermal Damage due to Preparation
Chemical- and ionic-thinning techniques					
Twin-jet electrolytic thinning			+		+
Full-bath electrolytic thinning			+		+
Twin-jet chemical thinning			+		+
Full-bath chemical thinning			+		+
Ion milling		+			+
Focused ion beam (FIB)		+			+
Mechanical-preparation techniques					
Grinding	+				
Wedge cleavage	+				
Tripod polishing	+				+
Ultramicrotomy	+				
Cryo-ultramicrotomy	+			+	+
Replica techniques					
Direct replica				+	
Indirect replica				+	
Extractive replica			+		
Freeze fracture	+			+	
Techniques specific to divided materials					
Fine particle dispersion				+	
Frozen hydrated film				+	
Contrast-enhancement and labeling techniques					
Decoration shadowing				+	+
"Negative-staining" contrast				+	
"Positive-staining" contrast			+		
Immunolabeling			+		

Bibliography

Ayache, J. and Albarede, P.H. (1994). *New Application of Klepeis Ion Mill Free Sample Preparation to YBCO System Thin Films and Bulk Ceramics,* P. H. ICEM 13, 1. Les Editions de Physique, Paris, 1023–1024.

Ayache, J. and Albarede, P.H. (1995). TEM cross section preparation by ionless tripod polisher method applied to YBCO superconducting multilayer thin films and bulk ceramics. *Ultramicroscopy*, **60,** 195–206.

Ayache, J., Thorel, A., Lesueur, J., and Dahmen, U. (1998). Characterization of a three-dimensional grain boundary topography in a YBa2Cu3O7 thin film bicrystal grown on a SrTiO3 substrate. *J. Appl. Phys.*, **84**(9), 4921–4928.

Ayache, J., Kisielowski, C., Kilaas, R., Passerieux, G., and Lartigue-Korinek, S. (2005). Determination of the atomic structure of a $\Sigma 13$ $SrTiO_3$ grain boundary. *J. Mater. Sci.*, **40,** 3091–3100.

Böhler, S. (1970). *Artefacts and Specimen preparationfaults in freeze etch technology*, Balzers AG. Liechtenstein.

Bouchet, D. and Colliex, C. (2003). Experimental study of ELNES at grain boundaries in alumina: Intergranular radiation damage effects on Al–L_{23} and O–K edges. *Ultramicroscopy*, **96,** 139–152.

Bouleau, D. (1974). La préparation des répliques destinées à l'observation des échantillons d'acier en MET-IRSID, 1996. Dalmai-Imelik G., Leclercq Ch., Mutin J., J. de *Microscopie*, **20,** 123.

Cavalier, A., Spehner, D., and Humbel, B.M. (eds.). (2008). *Handbook of Cryo-Preparation Methods for Electron Microscopy*. CRC Press, Boca Raton.

Crang, R.F.E. and Klomparens, K.L. (1988). *Artefacts in Biological Electron Microscopy*, Plenum Press, New York.

Kato, N.I., Kohno, Y., and Saka, H. (1999). Side-wall damage in a transmission electron microscopy specimen of crystalline Si pepared by focused ion beam etching. *J. Vac. Sci. Technol. A.*, **17**(4), 1201–1204.

McCaffrey, J.P., Phaneuf, M.W., Madsen, L.D. (2001). Surface damage formation during ion-beam thinning of samples for transmission électron microscopy. *Ultramicroscopy*, **87**, 97–104.

Chapter 7
Selection of Preparation Techniques Based on Material Problems and TEM Analyses

1 Introduction

The best choice of preparation technique is the one that produces a suitable thin slice of the material to be investigated. The technique must also be suitable for the different TEM analyses and should contain a minimum of artifacts.

Given the very small size of the sample to be analyzed in the TEM, the initial material will often need to be reduced using preliminary preparation techniques: sawing, cutting, dimpling, cleavage, etc. In order to conduct certain TEM analyses and to make observations in a precise direction, it is important to choose sample slices selected with specific orientations: longitudinal cuts, cross-sectional cuts, or cuts with a particular orientation. Sometimes it is necessary to combine several thinning techniques to optimize the TEM observation conditions.

First, one must have a good knowledge of the available techniques in order to select the one that will be most compatible with the material problem presented.

2 Classification of Preparation Techniques

Among the many preparation techniques for thin slices, a distinction is made between direct and indirect preparation methods.

Direct methods are used to analyze thinned materials or fine particles on a TEM support grid in order to perform structural, chemical, or spectroscopic analysis, i.e., to determine morphology, structure at different scales, crystallographic organization, chemical composition, and the nature of the chemical bonds. Most techniques in materials science are direct: twin-jet electrolytic thinning, full-bath electrolytic thinning, twin-jet chemical thinning, full-bath chemical thinning, focused ion beam (FIB) thinning, crushing, wedge cleavage, tripod polishing, ultramicrotomy and cryo-ultramicrotomy, and frozen hydrated film. Direct contrast enhancement techniques enable observation of the thinned materials or fine particles, but with the addition of chemical molecules or metallic particles in order to facilitate morphological investigation. These are positive- or negative-staining and decoration-shadowing techniques. They are not very suitable for chemical analyses.

J. Ayache et al., *Sample Preparation Handbook for Transmission Electron Microscopy*,
DOI 10.1007/978-0-387-98182-6_7,

Indirect methods do not enable direct observation of the material, as they involve the observation of an imprint of the material made using carbon films deposited on its surface, either with or without the shadowing effect of metallic particles. These techniques are used solely for analyses of the morphology of external surfaces or internal surfaces following fracture of the material. Included among them are direct- and indirect-replica and freeze-fracture techniques.

Techniques specific to bulk materials, multilayer materials, or fine particles (after embedding) are direct methods. Included among these methods are mechanical-, chemical-, electrochemical-, and ionic-thinning techniques, and the extractive-replica technique. The techniques most often used are tripod polishing, mechanical thinning, ion milling, and focused ion beam (FIB) thinning. They are essentially used in materials science. Only ultramicrotomy is used for biological materials as well.

These techniques, with the exception of the crushing and extractive-replica techniques, help to maintain the former microstructure and orientation of the material in the thinned slice.

Techniques specific to fine particles are direct methods. They are used in materials science as well as in biology. These techniques include fine particle dispersion and frozen hydrated films, as well as ultramicrotomy. The first two techniques result in random orientations of the particles in the thin slice. For ultramicrotomy, the orientation can either be random or not, and it depends on the shape of the particles (spheres, platelets, fibers, etc.) as well as on the possibility to orient them. All of these techniques enable statistical particle analysis, structure determination, and chemical analysis. For biological materials or polymers, it is often necessary to use contrast enhancement through the negative-staining or decoration-shadowing techniques.

Cryo-techniques make up a group of preparation techniques that involve cooling the material. They are mainly used in the investigation of biological materials, but are considerably developed for the preparation of thin slices of polymers in materials science. Cryo-ultramicrotomy is one of these techniques. Cryo-fixation techniques (often using cooled ethane) are mainly used in biology for investigating isolated particles and constitute an important part of the cryo-techniques. They are used to maintain the biological material in its hydrated structure.

3 Characteristics of Preparation Techniques

Each technique uses the material's mechanical, chemical, or electrical properties. The main characteristics of the thinning techniques are listed below:

Electrolytic thinning is used only on electrically conductive materials. The thin slices obtained do not contain any strain hardening.

Chemical thinning is used especially on non-electrically conductive materials. The thin slices obtained do not contain any strain hardening.

Ion bombardment (ion milling) can be used for any type of material except biological materials and soft polymers.

Focused ion beam (FIB) thinning is used to prepare all materials, including complex multiphase materials, while maintaining all phases and porosities, with the exception of soft biological materials and soft polymers, which have to be thinned using cryo-FIB.

The crushing technique can be used for any type of brittle material, but does not save sample microstructure.

The wedge-cleavage technique is used to prepare cross sections of thin or multilayer materials on a cleavable monocrystalline substrate.

The tripod polishing technique is used to prepare hard or brittle materials and also to obtain thin regions over large surface areas.

The ultramicrotomy technique is used to prepare any material, from soft to relatively hard, with very small dimensions. It is the main technique for biological materials and polymers and is also used for some minerals.

The cryo-ultramicrotomy technique is used to prepare any material that is very soft and/or has a liquid phase, over very small surface areas. This is the main technique for biological materials and polymers.

Replica techniques are used to prepare direct or indirect surface imprints of large-dimensional materials or isolated particles.

The freeze-fracture technique enables us to prepare, after fracturing the material, a topographical imprint of the internal surface of a biological material.

The fine particle dispersion technique is used to prepare a suspension of any material (bulk ground material, fine particles, or isolated material) of a thickness enabling TEM observation.

The frozen hydrated film technique is used to prepare a thin film of ice including isolated particles or macromolecules.

The decoration-shadowing contrast technique is used to enhance contrast using metallic particles.

The negative-staining contrast technique is used to enhance the contrast of the support of the organic fine particles via deposition of heavy-metal salts.

The positive-staining contrast technique is used to enhance the contrast of the thin slice of any organic fine particle via deposition of heavy-metal salts.

The immunolabeling technique is used to localize the functional sites of specific proteins.

4 Criteria Used to Select a Preparation Technique

The choice of a preparation technique for a thin slice depends on several criteria:

(i) Material type and properties
(ii) TEM analysis type
(iii) Specific preparation-induced artifacts
(iv) Thin slice orientation

All of the criteria involved in technique selection are shown in Fig. 7.1.

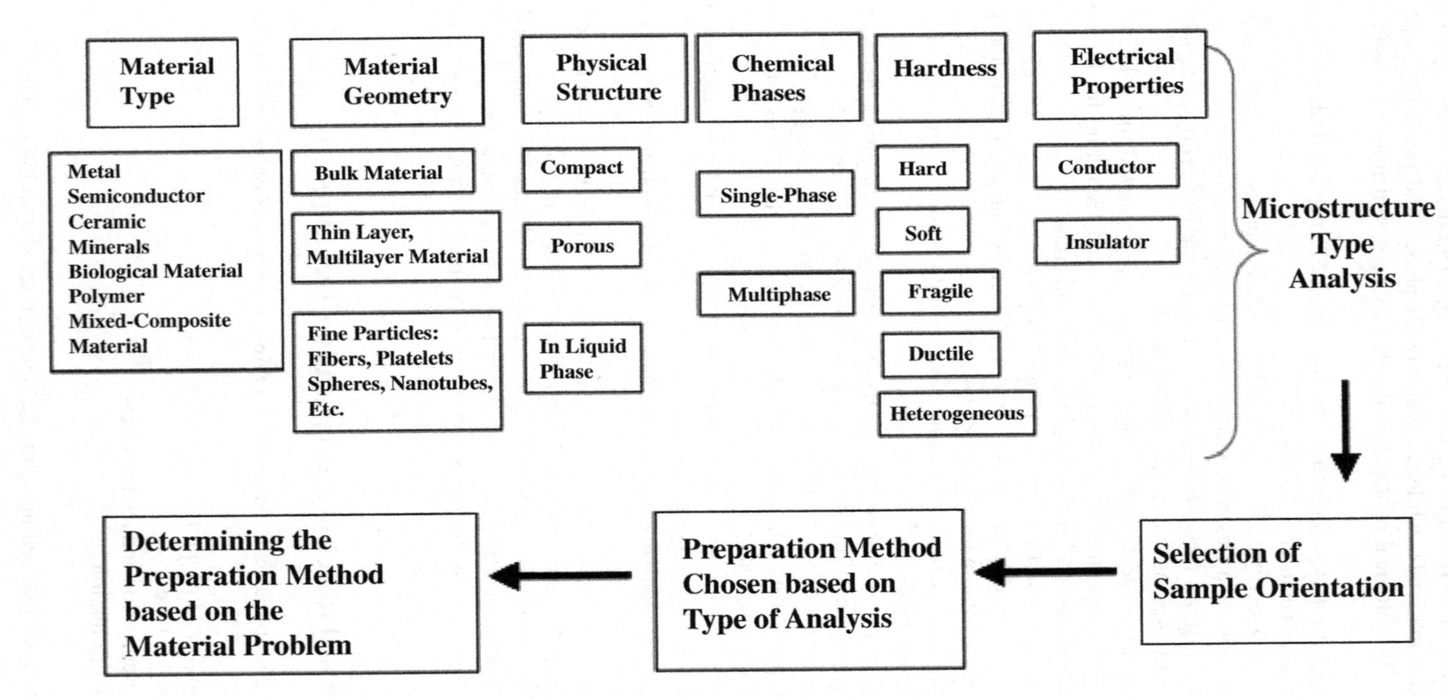

Fig. 7.1 Diagram of the approach to be taken for selecting a technique based on the material problem

5 Selection Criteria Based on Material Type

Metallic materials: Electrochemical and chemical preparations yield the best results. When the material is a single-phase material, electrolytic thinning is the most appropriate. If it is a multiphase material, chemical thinning is still possible, since a solution can be found for the different phases that results in a common dissolution. Otherwise, classic ion milling or FIB thinning is used.

Semiconductors: Electrochemical and chemical preparations yield the best results, but are applicable only if the material is composed of a single element such as silicon and more rarely for a multiphase semiconductors, for which it will be difficult to find a suitable solution to dissolve the different components. Mechanical thinning using tripod polishing, classic ion thinning, or FIB thinning is generally used.

Ceramics and minerals: For these two types of materials, the thinning techniques most often used are mechanical (crushing, tripod polishing, and ultramicrotomy) and ionic (bombardment and FIB). With the exception of some conductive ceramics, these materials are all insulators and electrochemical preparations do not apply. However, chemical thinning may be possible if the ceramic is a single phase and there is a known etchant.

Polymers: The preparation of a single-phase or multiphase polymer material can be performed using ultramicrotomy and cryo-ultramicrotomy, or possibly using gentle tripod mechanical thinning or cryo-FIB.

Biological materials: These materials are multiphase, hydrated, and unstable. Preliminary preparations for stabilizing the structure and increasing the hardness of the material are indispensable before the use of the ultracryomicrotomy technique. These prior chemical preparations make it possible to observe only dehydrated and embedded structures.

Conversely, cryo-techniques are used to study morphology in hydrated phases.

For biological materials such as viruses, bacteria, and isolated macromolecules, their structure is often observed using preparations of deposits of isolated particles at room temperature or in cryogenic mode (fine particle dispersion or frozen hydrated film).

Immunolabeling (cytochemistry) techniques are used for localizing proteins having a particular chemical function.

Composite or mixed–composite material: These are multiphase materials containing at least two crystalline and/or amorphous phases. This is the case with types of concrete, for example. For such a microstructure to be tied to the formation mechanisms and its properties, it must be observed while keeping the different phases together in the thin slice.

This type of material is difficult to prepare using classic ion bombardment, which creates much damage and deteriorates the interfaces. The best results are obtained using a soft mechanical tripod polishing technique, ultramicrotomy, or the focused ion beam (FIB) technique. These last two techniques help to keep the different phases together within the thin slice.

Polymers are often mixed multiphase materials. In order to investigate mixed polymer–metal or polymer–ceramic materials, etc., the classic ion bombardment technique cannot be used, since it would degrade all the interfaces. Depending on the material's hardness, if the second phase is a metal, semiconductor, ceramic, or any other blend, it may be prepared using tripod polishing techniques for hard materials and ultramicrotomy or cryo-ultramicrotomy for soft materials. The focused ion beam technique, either at room temperature or at low temperatures using cryo-FIB, can yield good results. This latter technique is not yet fully developed for TEM analyses on such materials, but it is still used for cryo-SEM observations.

6 Selection Criteria Based on Material Organization

6.1 Bulk Materials

In materials science, a bulk material does not present any particular difficulty and can be thinned, depending on its properties, using electrochemical, chemical, mechanical, or ionic thinning.

In biology, bulk material (tissue or cells) is usually hydrated and requires substitution–infiltration–embedding or freezing before being prepared in a thin section using ultramicrotomy or cryo-ultramicrotomy. In biology, there are also bulk materials that are hard and nearly dry, e.g., bone, tooth, wood, and chitin. These types of materials can either be thinned directly or following embedding if they are porous.

6.2 Single-Layer or Multilayer Materials

In materials science and biology, self-supporting single-layer materials are observable directly after being deposited on a TEM grid. In materials science, the materials composed of thin layers or multiple layers must be analyzed over areas large enough to be characterized. Depending on the material type, multilayer materials can be prepared using mechanical methods such as tripod polishing, ultramicrotomy, or ion thinning.

If the material is ductile, as with certain metals, ultramicrotomy can be used. In other cases, the tripod polishing technique, ion milling, and FIB thinning are often used. Tripod polishing is the technique of choice for semiconductor, ceramic, or composite materials. It is used to prepare large observable areas (sometimes up to a millimeter) and therefore provides information on several scales. In thin layer or multilayer materials prepared using ion milling, the thin areas are smaller.

In biology, single or multilayer most often correspond to cell cultures. We then come back to the problem of hydrated biological materials and to the ultramicrotomy or cryo-ultramicrotomy techniques, with the need to select the orientation of the slice: longitudinal or cross-sectional plane.

6.3 Fine Particles

In materials science and in biology, fine particles may be composed of fibers, platelets, powders, spheres, tubes, viruses, bacteria, or macromolecules. They can be observed either after being deposited on a TEM support film using the fine particle dispersion technique or embedded in a resin in order to make a block that can be handled and then sectioned using the ultramicrotomy technique. Some bulk materials can be made into fine particles after crushing (for materials in solid-state physics).

7 Selection Criteria Based on Material Properties

7.1 Based on the Physical State of the Material

Compact material: Does not present any particular difficulty, regardless of the technique used: electrochemical, chemical, mechanical, or ionic.

Porous materials: Depending on the size of the porosity, it will either be considered as a dense bulk material if the pore size is nanometric or as a porous material. In the latter case, it is important to fill the pores using infiltration before thinning the material. When the material is made bulk, it can be thinned using tripod polishing and ultramicrotomy mechanical techniques or the FIB technique. The best results will be obtained using a soft tripod polishing technique for hard materials, ultramicrotomy or cryo-ultramicrotomy for soft materials, or by the FIB technique, enabling us to maintain the internal porosities. For porous biological materials, it is necessary to replace the air contained in the pores with water or a solvent, as is the general case for hydrated materials.

Materials with a liquid solution (mainly biological materials): This type of material concerns all bulk or dispersed biological materials. The preparation of bulk materials will require the elimination of water through dehydration, substitution using a solvent, and then using an embedding resin before using the ultramicrotomy technique. One can also transform the water into solid phase in order to use the cryo-ultramicrotomy or the freeze-fracture technique.

When the material is composed of fine particles in aqueous phase, it can contain viruses, bacteria, or macromolecules (DNA, proteins, etc.). It can be observed either directly after deposition onto a TEM support film or after the carbon support film is treated. This treatment is positive ionization for depositing DNA and a negative ionization for proteins. These fine particles can also be prepared using the frozen-hydrated film technique and can be observed using cryo-TEM after cryo-transfer.

7.2 Based on the Chemical Phases in the Material

Knowing the chemical nature of the material, in particular the type of chemical bonds and its chemical stability (from its phase diagram), helps to predict the secondary phases that can be formed during chemical, ionic, and physical actions. Depending on whether or not the material contains one or more types of chemical bonds, it will be affected more or less significantly by chemical reactions that might occur.

Materials with metallic bonds are sensitive to mechanical actions.

Materials with covalent bonds only are stable and will not be disturbed (or only slightly) by chemical or ionic effects. We find them in semiconductors, some ceramics, some polymers, as well as some composites.

Materials with ionic bonds only will be sensitive to ionic and electronic radiation. Some ceramics and minerals are included here.

Materials with multiple bonds contain several types of bonds. They coexist in the materials of solid-state physics and biological materials, giving them completely individual properties. In fact, the presence of a mixed valence, i.e., mixed ionic and covalent bonds carried by a type of atom, makes the material more sensitive to non-stoichiometry. This is the case with many ceramic materials (superconductors, ferroelectrics, ferromagnetics, magneto-resistant materials, etc.), some minerals, and composite or mixed–composite materials. The effects of ion milling on these materials can result in variations in stoichiometry and can transform the physical properties. In the case of biological materials, with the exception of metallic bonds, they contain all types of strong and weak chemical bonds. The difficulty with these materials also derives from the fact that the constituents of the microstructure are themselves complex and in an aqueous medium, rendering them very brittle during preparation and observation.

7.3 Based on the Electrical Properties of the Material

Electrochemical and chemical preparations yield the best results for materials presenting electrical properties. In order to perform final thinning using electropolishing, it is mandatory to have a polarizable material, and therefore an electrical conductor, however weak, as with certain semiconductors. For chemical polishing, the material must be able to take part in a chemical reaction by means of natural polarization in the chemical bath. For single-phase materials, electrolytic thinning is the most suitable. If it is a multiphase material, chemical thinning is still possible when a common etchant can be found for the different phases.

7.4 Based on the Mechanical Properties of the Material

7.4.1 Materials in Solid-State Physics

The mechanical properties of the material to be thinned will define a range of unavoidable constraints based on the type of structural analysis selected for use, in particular for crystal defect analysis. Any preparation involving mechanical polishing can induce strain-hardening defects of the material, and extrinsic dislocations are all the more significant depending on whether the material is ductile, soft, or resistant. These are mechanical-thinning techniques that will present problems: either the material is hard and brittle and sample breakage may occur or the material is soft and strain hardening is likely. Between these two extremes, the technique must be adapted to preserve the intrinsic structure of the material.

7.4.2 Soft-Ductile Materials

The softer a material, the more its structure will be sensitive to mechanical pressures during preparation. Thus, the investigation will ruin a liquid crystal or mesophase carbon structure if the mechanical-grinding technique is used, whereas the material will be saved if the ultramicrotomy technique is used, as this technique minimizes mechanical damage. Generally speaking, the best preparation technique for bulk and fine particle soft samples is ultramicrotomy or cryo-ultramicrotomy. For bulk soft materials, different thinning techniques can be used: chemical, electrochemical, ionic, and FIB thinning. If the sample is a thin layer or multilayer material, the best result should be obtained using ultramicrotomy or cryo-ultramicrotomy, or mechanical thinning followed by ionic or short chemical thinning.

7.4.3 Hard-Resistant Materials

For hard materials, ultramicrotomy will be ineffective because interfaces are not kept complete due to the hardness of the material confronted with the hardness of the diamond knife. In this case, mechanical polishing will be more favorable because it helps to keep grain boundaries and interfaces together, even down to small thicknesses.

For single-phase and hard materials, mechanical polishing combined with ionic thinning can produce good results. Likewise, in this case, FIB ionic and electrochemical polishing also produces good results.

For multiphase materials, only gentle mechanical polishing can help to obtain large observable areas, while maintaining all of the phases. In this case, FIB produces very good results.

Preparing hard samples by means of mechanical polishing does not present any particular limitation. However, the granular nature of a ceramic or composite can result in a minimum critical thickness of the sample in order for it to maintain its mechanical cohesion. If this thickness is not thin enough to be electron transparent, final ionic or chemical thinning may then be necessary.

7.4.4 Materials of Intermediate Hardness and Ductility

For intermediate cases, pure mechanical thinning should be tried. If this is not possible, thickness should be reduced as much as possible with gentle tripod polishing and a final ionic, chemical, or electrochemical thinning for just a few minutes, so as to maintain the interfaces between phases or matrix and precipitates. The FIB technique produces very good results.

As in the previous case, for materials containing a mix of soft and hard phases, it is necessary to reach the sample's critical thickness and end up performing an ionic or chemical thinning. FIB thinning again produces very good results.

The preparation method for obtaining thin slices must be adapted based on the material's hardness, i.e., based on its mechanical behavior or the mechanical behavior of the various phases.

7.4.5 Biological Materials

With the exception of hard materials such as bone, tooth, wood, and chitin, most biological materials are very soft, since they are in liquid phase. They are hardened by using the infiltration–embedding technique, so they can be prepared using ultramicrotomy and the freezing technique for cryo-ultramicrotomy.

Table 7.1 lists the different types of materials, as well as the possible preparation techniques, depending on their nature. Note that metals and alloys, considered to be soft in comparison with ceramics, are often investigated for their mechanical

Table 7.1 Examples of materials of different hardnesses and indication of the possible preparation techniques

Soft-ductile materials	Intermediate materials	Hard-resistant materials
Biological materials (living cells, living organic matter) Polymers (epoxy, phenolic, acrylic, resins, etc.) Liquid crystals (pure carbon with nematic stacking) Some reinforced polymer composites Organic, natural, and artificial matter (mesophase carbon, carbon or petroleum tars, kerogenes) Metals, "bulk alloys" Thin layers Metallic multilayer materials on metallic or semiconductor substrates	Semiconductors Metal/ceramic composites: mesophase type carbon Composites: carbon fiber/polymer matrix Multiphase metal/oxide/semiconductor materials in thin layers (electronic components) Concrete: oxide–polymer mixtures Thin metallic layers on oxide substrate	Simple oxides (alumina) Ceramic materials (carbides, complex oxides, mixtures of carbides and oxides) Composites (carbon/carbon composites, ceramic/ceramic deposits) Thin oxide layers on an oxide substrate Doped alloys
Mechanical preparation using microtomy and ultramicrotomy, chemical, electrochemical thinning, FIB	**Tripod polishing, or tripod + chemical, or tripod + ion milling, electrochemical thinning, FIB**	**Tripod polishing, tripod + ion milling, tripod + chemical etching, FIB**

properties. They are often prepared using mainly chemical or electrochemical methods in order to prevent any mechanical-type artifacts.

8 Selection Criteria Related to the Type of TEM Analysis

The type of analysis to be performed will condition the selection of one preparation over another. Selection criteria are ultimately related to the following question that must be answered: Is it an investigation of topography, structure, structural defects, crystallography, chemical composition, or chemical bonds or is it an investigation of physical, chemical, and functional properties of the material?

Topography Investigation: When a bulk material cannot be thinned, replicas of its surface can be made using direct- or indirect-replica techniques in order to investigate its topography using CTEM. For hydrated organic materials, the internal topography of the material can be investigated using freeze fracture.

Structural Analysis: *Organization and Structure, Interface, and Volume: To investigate microstructure, a sample preparation technique must be used that allows all the components to remain in place until the thin slice is made.* Mechanical thinning techniques such as tripod polishing, ultramicrotomy, ion bombardment, and FIB make this possible. The investigation of a single crystal atomic structure can be done using the simplest technique, e.g., crushing, which will yield a statistic of its crystal projections. In biology, only the ultramicrotomy technique can be used to obtain a thin slice.

Crystallographic Analysis: Crystallographic analyses can be made following electrochemical, mechanical, and ionic preparations. However, the best diffraction patterns are obtained on samples prepared with mechanical preparations such as tripod polishing, ultramicrotomy, or FIB thinning, due to the constant thickness of the slice. Under these conditions, qualitative analysis of the stresses at the interfaces can be performed and combined with the crystallographic structure and possible quantitative chemical analyses.

The material can have an orientation dictated by its processing. Depending on the need to investigate longitudinal plane orientation in cross section with regard to the preferential orientation or a particular crystallographic direction, bulk samples have to be pre-oriented; fine particles must be pre-embedded before thinning. This problem is not present for fine particles, where the orientation is random. As a consequence, due to the statistical distribution of all of the particular orientations on the support film, crystallographic analysis is possible.

Crystal Defect Analysis: The investigation of crystal defects dictates a particular orientation of the thin slice so that the defects can be oriented. In the case of thin films or heterostructures, the full structural analysis of the interfaces and growth defects requires the preparation of a longitudinal-plane section in order to identify the interfaces in the film plane and a cross section to identify the interfaces between the layers or with the substrate. Therefore, in the case of a ceramic, a grain boundary or interface analysis will require several observations of these defects with different orientations. Therefore, several preparations with different orientations will be necessary. The same is required for the analysis of dislocations.

The electrochemical, chemical, and ionic preparation techniques are used to investigate structural defects. Mechanical preparation techniques generate strain hardening and are not favorable to the investigation of dislocations.

Chemical Analysis: *Phase Identification and Concentration Profiles*: Chemical composition analysis of a bulk material, thin layer material, or fine particle material requires the use of a preparation technique that does not result in chemical changes induced by the technique. The best preparation techniques are mechanical techniques such as tripod polishing for bulk and multilayer materials, wedge cleavage for cleavable multilayer materials, and ultramicrotomy for fine particles after embedding. The ionic preparation techniques can also be used if there is no modification in the chemical composition of the material under the ion effect.

Chemical analysis of interfaces can be performed on samples prepared using the tripod and ultramicrotomy techniques and ionic techniques. However, during ion bombardment, the ionic differential sputtering rates will favor certain grain orientations and result in significant thickness variations along the grain boundary. The quantitative chemical analysis of these boundaries will then be more difficult. Preparations using FIB or tripod polishing help to solve this problem. The FIB technique is used to obtain a thin slice with a constant thickness, but with the drawback of surface amorphization. The advantage of the tripod polishing method is that it does not produce surface amorphization caused by ionic methods, which are very unfavorable in HRTEM analysis.

Chemical Bond Analysis: The electrochemical, chemical, mechanical, and ionic preparation techniques are used to prepare samples for EELS analyses. However, the specimen must have a thickness less than 50 nm.

Physical Property Analysis: Investigations of mechanical and electrical properties and in situ investigations under high temperature require a sample that is not too thin along the edges and that is mechanically resistant in order to undergo stresses during observation without breaking. The main preparation technique is ion milling. To investigate the optical or electronic properties using EELS analyses, mechanical and ionic techniques are possible.

Functional Analysis: The immunolabeling technique, which is specific to biology, helps to localize functional sites or specific molecules.

8.1 Preparation Techniques

The table below provides cross-references for the numbers used in Tables 7.2, 7.3, and 7.4 and the preparation number provided in the "Techniques".

1 = 3.1 and 3.2	Electrolytic thinning
2 = 3.3 and 3.4	Chemical thinning
3 = 3.5	Ion milling
4 = 3.6	Focused ion beam thinning (FIB)
5 = 4.1	Crushing
6 = 4.2	Wedge cleavage
7 = 4.3	Tripod polishing
8 = 4.4 and 4.5	Ultramicrotomy and cryo-ultramicrotomy
9 = 5.1	Direct replica
10 = 5.2	Indirect replica
11 = 5.3	Extractive replica
12 = 5.4	Freeze fracture
13 = 6.1	Fine particle dispersion
14 = 6.2	Frozen hydrated film
15 = 7.1	Decoration shadowing
16 = 7.2	"Negative-staining" contrast
17 = 7.3	"Positive-staining" contrast
18 = 7.4	Immunolabeling

Tables 7.2, 7.3, and 7.4 list the different suitable preparation techniques for bulk and multilayer materials and fine particles based on the type of analysis to be performed.

9 Selection of the Orientation of the Sample Section

Crystallographic and structural material investigation requires one or more specific orientations of the sample, based on the original material and with regard to the direction of observation in the microscope. For this, the sample can be tilted in the microscope up to maximum angles of 60° (depending on the type of microscope) or can be orientated beforehand when it is prepared. Orientation is determined with regard to the geometry of the thin slice to be observed in the TEM. In fact, an unavoidable aspect of transmission electron microscopy is the changeover from 3D of the volumic sample to the 2D projection in the image (Fig. 7.2). The changeover results in the superimposition of information contained in a volume (the thickness of the sample) to a plane. The direction of projection, i.e., the thickness of the sample combined with the direction of electron propagation, is therefore singular. Consequently, it is important to take into account the geometry of the structures observed with regard to this direction.

Specimens can be observed in different orientations by preparing longitudinal slices, cross-sectional slices, or in a precisely defined direction, e.g., parallel to a set of atomic planes. Lastly, the sample can be observed in any direction, i.e., based on a non-specific orientation or in random directions (Table 7.5).

Table 7.2 Selection of the preparation technique based on the material and the type of analysis: Bulk materials

Material	Metal	Semiconductor	Ceramic	Mineral	Polymer	Biological material	Mixed–composite material
Type of analysis							
Topography	9, 10	9, 10	9, 10	9, 10	9, 12	9, 12	9, 10
Structure, organization	1, 2, 3, 4, 5, 8, 11	1, 2, 3, 4, 5, 6, 7	2, 3, 4, 5, 7, 11	2, 3, 4, 5, 7, 8, 11	8, 12, 15	8, 12, 17, 18	3, 4, 7, 5, 8, 11
Crystal defects	1, 2, 3, 4, 11	1, 2, 3, 4, 5, 6, 7	2, 3, 4, 5, 7, 11	2, 3, 4, 5, 7, 11			3, 4, 7, 5, 11
Crystallography	1, 2, 3, 4, 8, 11	1, 2, 3, 4, 5, 6, 7	2, 3, 4, 5, 7, 11	2, 3, 4, 5, 7, 8, 11		8	3, 4, 7, 5, 8, 11
Chemical composition	1, 2, 3, 4, 8, 11	1, 2, 3, 4, 5, 6, 7, 8	2, 3, 4, 5, 7, 11	2, 3, 4, 5, 7, 8, 11	8	8	3, 4, 7, 5, 8, 11
Chemical bonds	1, 2, 3, 4, 8, 11	1, 2, 3, 4, 5, 6, 7, 8	2, 3, 4, 5, 7, 11	2, 3, 4, 5, 7, 8, 11	8	8	3, 4, 7, 5, 8, 11
Properties	1, 2, 3, 4, 8, 11	1, 2, 3, 4, 5, 6, 7, 8	2, 3, 4, 5, 7, 11	2, 3, 4, 5, 7, 8, 11	8, 17	8, 12, 17, 18	3, 4, 7, 5, 8, 11

Table 7.3 Selection of the preparation technique based on the material and the type of analysis: thin layer and multilayer materials

Material	Metal	Semiconductor	Ceramic	Mineral	Polymer	Biological material	Composite
Type of analysis							
Topography	9, 10	9, 10	9, 10	9, 10	9	9	
Structure, organization	1, 2, 4, 8	1, 2, 3, 4, 6, 7, 8	3, 4, 5, 7, 8	3, 4, 5, 7, 8	8	8, 17, 18	3, 4, 7, 8
Crystal defects	1, 2, 3, 4	1, 2, 3, 4, 6, 7	3, 4, 5, 7	3, 4, 5, 7			3, 4, 7
Crystallography	1, 2, 3, 4, 8	1, 2, 3, 4, 6, 7, 8	3, 4, 5, 7, 8	3, 4, 5, 7, 8			3, 4, 7, 8
Chemical composition	1, 2, 3, 4, 8	1, 2, 3, 4, 6, 7, 8	3, 4, 5, 7, 8	3, 4, 5, 7, 8		8	3, 4, 5, 7, 8
Chemical bonds	1, 2, 3, 4, 8	1, 2, 3, 4, 6, 7, 8	3, 4, 5, 7, 8	3, 4, 5, 7, 8	8	8	3, 4, 5, 7, 8
Properties	1, 2, 3, 4, 8	1, 2, 3, 4, 6, 7, 8	3, 4, 5, 7, 8	3, 4, 5, 7, 8	8, 17	8, 17, 18	3, 4, 5, 7, 8

Table 7.4 Selection of the preparation technique based on the material and the type of analysis: Fine particles

Material	Metal	Semiconductor	Ceramic	Mineral	Polymer	Biological material	Composite
Type of analysis							
Topography					9, 12, 13, 15, 16	12, 13, 15, 16	
Structure, organization	3, 8, 11, 13	3, 5	3, 5, 8, 11, 13	3, 5, 8, 11, 13	8, 12, 13, 14, 15, 16, 17	8, 12, 13, 14, 15, 16, 17, 18	3, 5, 8, 11, 13
Crystal defects	3, 11, 13	3, 5	3, 5, 11, 13	3, 5, 11, 13			3, 5, 11, 13
Crystallography	3, 8, 11, 13	3, 5	3, 5, 8, 11, 13	3, 5, 8, 11, 13			3, 5, 8, 11, 13
Chemical composition	3, 8, 11, 13	3, 5	3, 5, 8, 11, 13	3, 5, 8, 11, 13	8, 13	8, 13	3, 5, 8, 11, 13
Chemical bonds	3, 8, 11, 13	3, 5	3, 5, 8, 11, 13	3, 5, 8, 11, 13	8, 13	8, 13	3, 5, 8, 11, 13
Properties	3, 8, 11, 13	3, 5	3, 5, 8, 11, 13	3, 5, 8, 11, 13	8, 12, 13, 16, 17, 18	8, 12, 13, 16, 17, 18	3, 5, 8, 11, 13

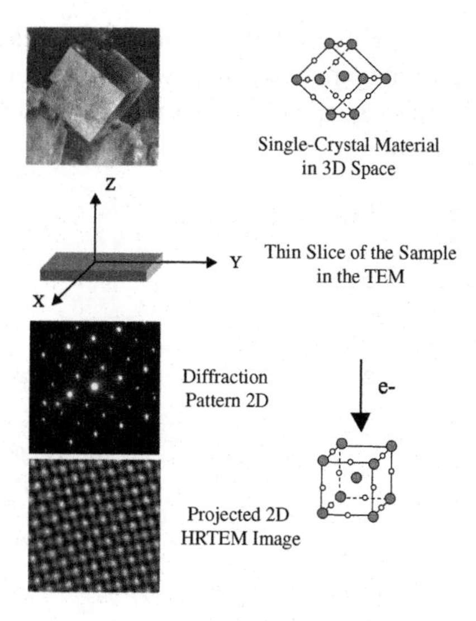

Fig. 7.2 Three-dimensional material and the 2D projection of the thin slice in the TEM observed in diffraction and HRTEM modes

Table 7.5 Possible cutting directions and sections based on the nature of the material

Bulk material without preferential orientation in the block of material	**Massive textured bulk material with preferential orientations**
Any cut in the volume of the self-supporting bulk material Cut in any direction	Longitudinal plane cut parallel to the preferential orientation or the substrate Cross section: perpendicular to the preferential direction or to the substrate Particular cut based on orientation Any direction
Thin layer or multilayer material	***Fine particles***
Longitudinal plane cut parallel to the layer Longitudinal plane cut parallel to the substrate Cross section: perpendicular to the preferential direction of the layers or to the substrate–film interface	Longitudinal plane parallel to the preferential orientation of the particles Cross section: perpendicular to the preferential direction of the particles Any cut of the particles Random cut of the particles

9.1 Microstructure Geometry

A monocrystalline material (single crystal and thin film) can be observed based on certain crystallographic directions, i.e., directions that are either related to the growing direction, determined beforehand by the orientation of the material itself, or are based on certain particular atomic planes with known orientation. In these cases, a longitudinal plane, cross section, or particular orientation can be selected.

For polycrystalline materials, the challenge lies in investigating the structure of special grain boundaries with particular crystallographic orientations in a cluster of grains having any orientation. The observation direction of the sample can be in any direction (Fig. 7.3).

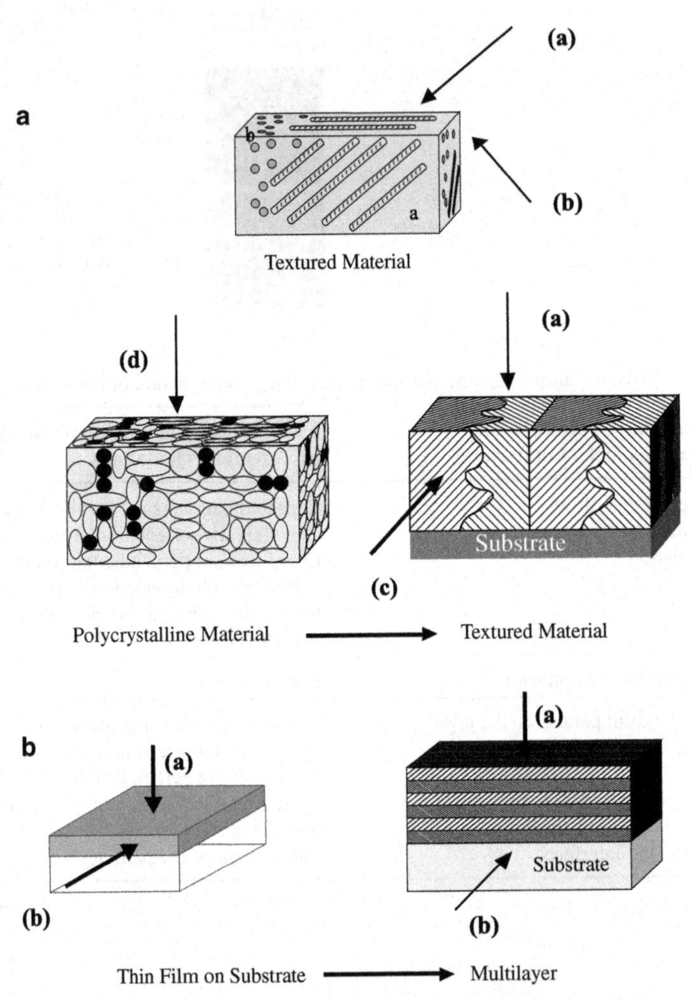

Fig. 7.3 **a**) Different cutting directions of a bulk textured or polycrystalline material in order to make thin slices: (*a*) longitudinal plane cut parallel to a surface, or fiber direction, or main direction (e.g., as in a texture); (*b*) cross section perpendicular to a main fiber direction or an interface; (*c*) particular cut with regard to the orientation of a given set of atomic planes, an interface, a grain boundary, etc., and (*d*) any cut, such as in a polycrystalline material with no preferential orientation. **b**) Different cutting directions of single-layer and multilayer materials in order to make thin slices: (*a*) longitudinal plane parallel to a surface of the substrate or a film and (*b*) cross section perpendicular to the substrate – thin layer, layer–layer, or multilayer interface

In the case of a textured or multilayer material, it is possible to observe the microstructure in a thin slice in the direction parallel to the texture (longitudinal plane) or in the perpendicular direction (cross section).

In the case of fine particles of various shapes, it will be possible to observe all orientations due to their random deposition. Figure 7.4 shows some 2D projections of particles based on their shapes and orientation on the slice.

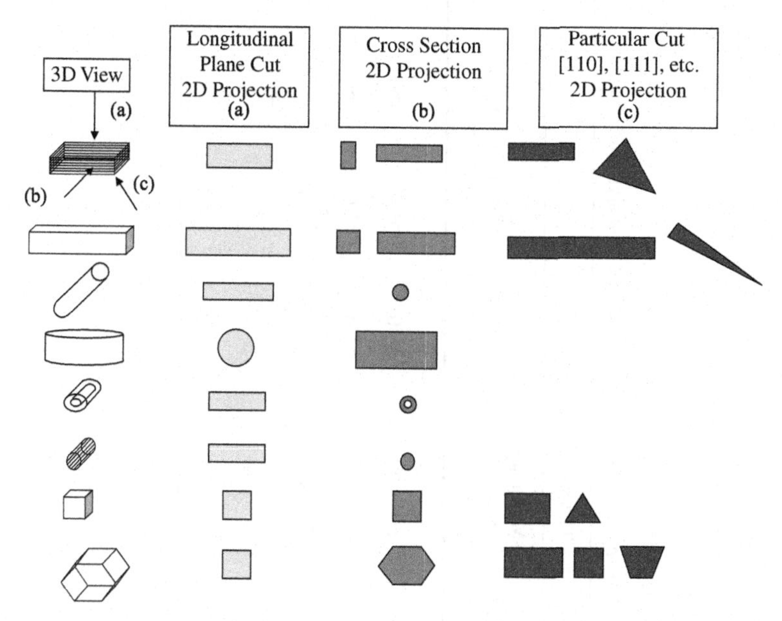

Fig. 7.4 Fine isolated particles of varying form in 3D view and 2D projections: (*a*) longitudinal plane cut; (*b*) cross-sectional cut; and (*c*) a cut specific to a crystallographic orientation

9.2 Defect Geometry

The characterization of crystal defects requires a particular orientation based on their dimensions. The analytical techniques and their limitations are relatively easy to use based on the thickness of the samples, the geometry of the defect in the sample, and the size and density of the defects. In all cases, the problems of projection and superimposition must be taken into account based on the selected TEM technique.

Techniques for preparing thin slices by electrolytic, chemical, mechanical, or ionic thinning do not systematically result in directly observable thin slices for all of the analytical techniques. For example, the observation conditions for the same dislocation in weak-beam and high-resolution modes are hardly compatible. The same goes for combining CBED-LACBED diffraction and HREM imaging modes, which do not require the same sample thickness. Table 7.6 shows the slice thickness required, depending on the observation mode and the analysis to be performed.

Table 7.6 Slice thickness based on the observation mode and type of analysis

Observation mode	CTEM				TEM/STEM and STEM					
Analysis type	BF/DF (nm)	WB (nm)	SAED (nm)	HRTEM (nm)	Micro- and nano-diffraction (nm)	CBED (nm)	LACBED (nm)	*HAADF* (nm)	EDS (nm)	EELS (nm)
Topography	<50									
Structure				<50	<50	≥50	≥50	<50		
Structural defects	<50	≥50	≤50	<50	<50	≥50	≥50	<50		
Crystallography			≤50		≤50	≥50	≥50			
Chemical composition								≤50	≥50	<50
Chemical bonds										<50
Properties	≥50			<50		≥50	≥100			<50

10 Selection Criteria Related to Artifacts Induced by the Preparation Technique

Each technique causes particular artifacts. It is necessary to choose the technique whose induced artifacts present the least amount of problems for the analysis planned. For example, to investigate the mechanical properties of a material, any mechanical preparation must be avoided that produces extrinsic defects that would be superimposed over the intrinsic defects. Therefore, it is very important to know the artifacts induced by a specific technique (see Chapter 6).

There are four classes of artifacts corresponding to the four types of preparation techniques: mechanical, chemical, ionic, and physical. These classes are, in addition to thermal defects, caused during thinning. For multiphase materials containing phases with very different mechanical properties, ultramicrotomy preparation will reveal these differences and extract the hard phases. If the material is hard, the tripod polishing technique is suitable. If the material is of average hardness, an ionic thinning technique (e.g., FIB) will be preferred since it keeps all of the phases, even if they have mechanically different properties, and does not introduce dislocations. In some cases, there is no alternative to this technique: e.g., biological materials prepared primarily using ultramicrotomy. Specific defects must be recognized and the observation must be performed on the structure that has not been deformed by the knife during cutting. The major artifact resulting from dehydration must be avoided in order to study the spatial conformation of proteins. Only cryo-techniques can be used to observe them in the hydrated state.

In all cases it is necessary to select the technique inducing the fewest artifacts in the material to be studied (see Chapter 6). The results of several types of preparation techniques should be compared in order to ensure that the true nature of the structure is observed.

11 Adaptation of the Technique Based on Problems Related to Observation

Sometimes one technique alone is not enough to thin a sample down to electron transparency. In other cases, it forms artifacts that can be eliminated in order to optimize the observation conditions by combining several thinning techniques. Improvements to TEM observation can be made by reducing the thickness of the thin slice or by eliminating the contamination or amorphization layers using chemical or ionic etching. Poor contrast may be increased by adding heavy elements.

11.1 Reducing Sample Thickness

During chemical, ionic, or tripod polish thinning, the same thickness cannot be obtained on both sides of an interface when the material presents phases of different compositions. This is due to the differential thinning rates in these techniques.

The origin of this problem may be linked to the nature of the etchant or to the presence of heavy atoms in the material that are difficult to polish. The way to resolve this problem is to complete the thinning of the thin slice using ionic thinning following chemical thinning, or by using a short chemical thinning after the ionic or mechanical thinning.

11.2 Increasing Contrast

For materials composed of light atoms, e.g., polymers or biological materials, contrast can be increased by adding contrasting agents using the "positive-staining" contrast technique. This will reveal the different phases.

For fine particles with a low atomic number, contrast can be increased using "negative staining" or through shadowing. These techniques highlight the reliefs and surface-morphology details.

11.3 Reducing Charge Effects

During TEM observation, if the charge dissipation is insufficient, as is the case for insulating materials, the sample becomes unstable under the beam and observation is not possible. The way to solve this problem is to coat the sample with a very thin carbon film in order to eliminate these effects.

11.4 Limitation of Strain Hardening

For brittle or ductile materials, thinning techniques can result in strain hardening of the sample. The way to limit strain hardening is to carry out a final ion milling or FIB thinning, or chemical or electrochemical etching.

11.5 Removal of Surface Amorphization

Ion milling, and to a lesser extent FIB, creates amorphization of the surface layers, which can reach a few nanometers and interfere with HRTEM observations. These amorphous surface layers can be reduced by brief chemical etching; this is possible principally for single-phase materials.

11.6 Removal of Surface Contamination

Following the electrolytic and electrochemical thinning preparations, a layer of surface contamination may form. It can be eliminated after a final brief ion milling process at low energy (100–500 eV, low-angle ions, sample cooled down with liquid nitrogen).

11.7 Final Cleaning of the Thin Slice

Samples that are stored for more than a few hours will be contaminated by the atmosphere. Before observation, and especially before HRTEM observation or microanalysis, it is essential to clean the sample surfaces in order to prevent any additional contamination under the electron beam. This step can be carried out using a plasma cleaner (argon, argon/oxygen, etc.). This step is necessary for observations with a FEG microscope (TEM/STEM and STEM).

12 Conclusion

Depending on the type of analysis to be made, the material type, and the material properties, we generally can choose from several techniques.

Table 7.7 Flowchart of the possible methods for preparing different types of thin slices based on the sample types

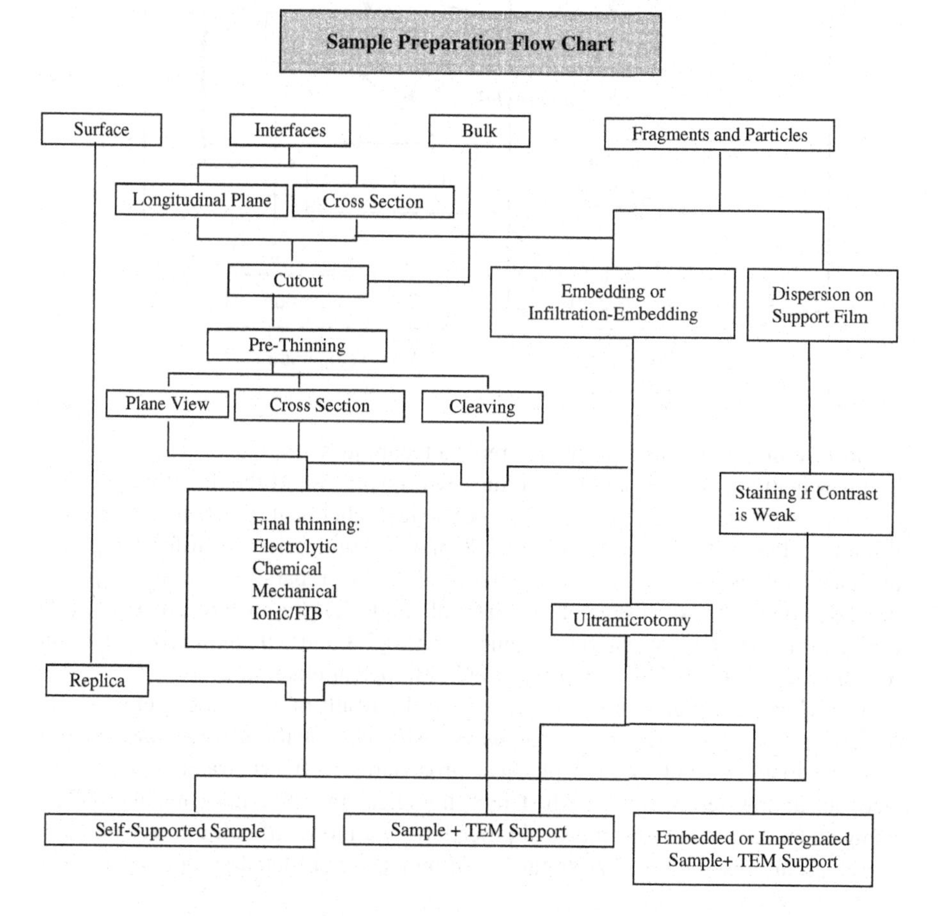

Table 7.8 Selection of the preparation technique for bulk biological materials

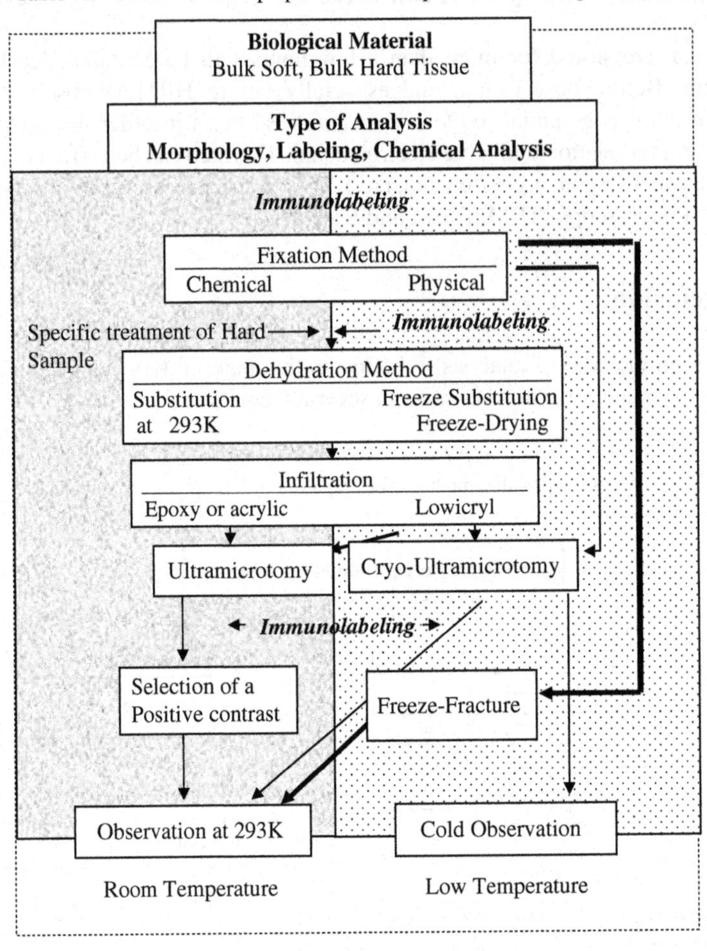

Surface investigations will involve replica techniques.

Investigations of interfaces and volume will involve all of the direct preparation techniques for materials science (electrochemical, chemical, mechanical, or ionic thinning). Tables 7.2, 7.3, 7.4, 7.6, 7.7, 7.8, and 7.9 summarize the different options available. Tables 7.2, 7.3, and 7.4 summarize all of the preparation techniques possible based on the physical state of the material. Table 7.7 presents the most common options for materials in materials science. Tables 7.8 and 7.9 summarize the main techniques possible for fine particles or bulk biological materials.

In biology, the preparation techniques are the result of a complex set of steps. There are two major preparation pathways: using chemical reactions that occur at room temperature and through physical procedures to freeze water that are carried out at low temperatures. Most morphological investigations are undertaken using chemical procedures; physical procedures are more difficult to carry out and, therefore, are reserved for investigations of molecules (isolated particles) and their

Table 7.9 Selection of the preparation technique based on the material and analysis type: biological fine particles

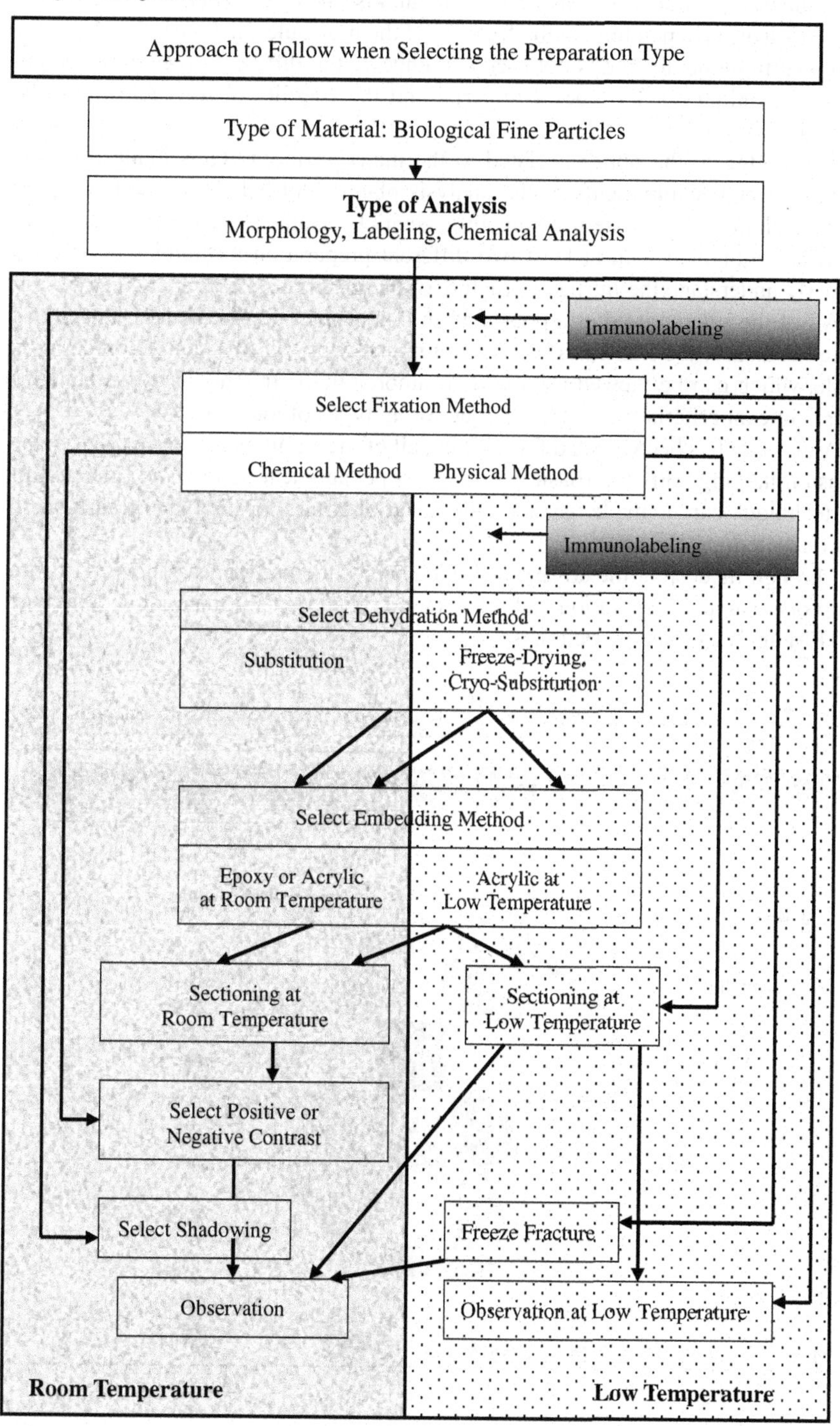

properties (immunolabeling on tissues, for example). These two ways of investigating a material can be used separately, but can also be changed at any time during the investigation depending on the fragility of the molecules that are to be observed. Note that immunolabeling is not used just once, but at different times in the preparation depending on the type of material and the fragility of the molecule to be labeled.

Knowledge of the options related to the material, the artifacts induced by the technique, and requirements of the analysis planned all help to refine the choice of the technique to be used. It remains essential, insofar as possible, to compare the results obtained with at least two different preparation techniques in order to properly understand the intrinsic structure of the material.

All of the selection criteria presented in this chapter are listed in an interactive online database, the transmission electron microscopy (TEM): sample preparation guide, which can be viewed at http://temsamprep.in2p3.fr. This database contains the significant criteria for each technique for all types of materials.

A glossary has been created for defining all of the significant criteria used. In the decision-helping software (basic guide and a pedagogical guide), the selection of the technique can be made based on the material criteria or the type of analysis to be conducted.

Through its design, the advanced guide enables the user to learn how to choose the proper sample preparation technique based on the type of material and the type of TEM analysis to be carried out.

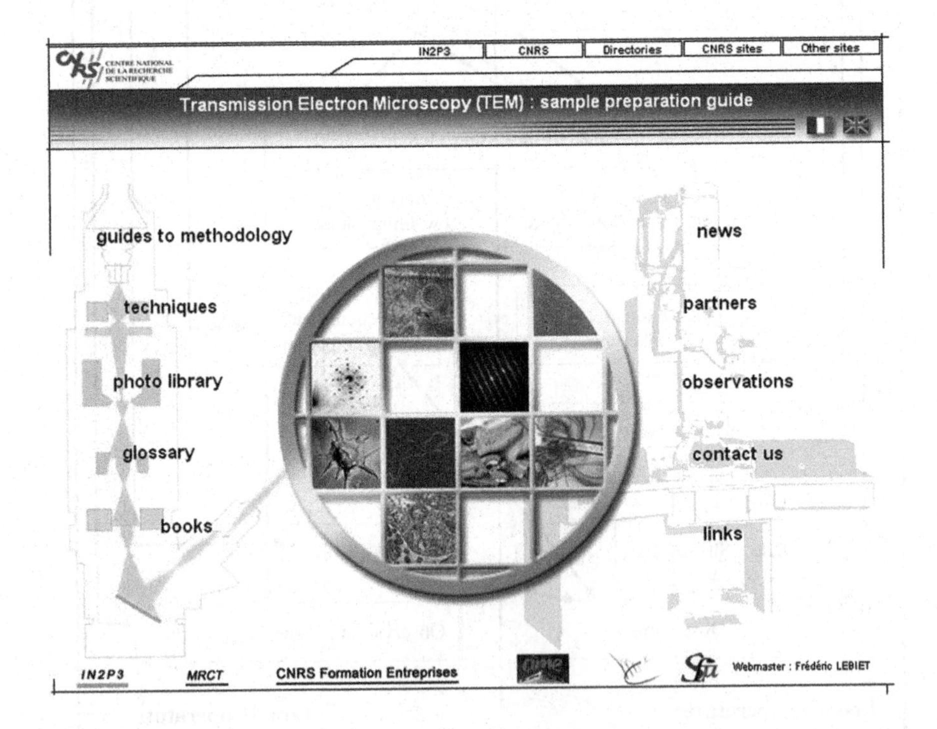

There are 36 sample preparation techniques listed, and their main characteristics are described. These 36 techniques are developed in "Techniques" of this book. An interactive image gallery presents the different types of materials prepared using these techniques and also provides examples of materials and techniques compiled by the international community of microscopists.
http://temsamprep.in2p3.fr

Bibliography

Anderson, R.M. (1990). *Specimen Preparation for Transmission Electron Microscopy of Materials II*, MRS Matérial Research Society Symposium Proceedings, vol. 199, Pittsburgh, PA.

Anderson, R.M., Tracy, B., and Bravman, J. (1992). *Specimen Preparation for Transmission Electron Microscopy of Materials III*, MRS Matérial Research Society Symposium Proceedings, vol. 254, Pittsburgh, PA.

Bousfield, B. (1992). *Surface Preparation and Microscopy of Materials.* Wiley and Sons, Buehler Europe Ltd, Coventry.

Bravman, J., Anderson, R.M., and Mcdonald, M.L. (1988). *Specimen Preparation for Transmission Electron Microscopy of Materials*, MRS Matérial Research Society Symposium Proceedings, vol. 115, Pittsburgh, PA.

Delain, E. and Le Cam, E. (1995). The spreading of nucleic acids. In *Visualization of Nucleic Acids,* vol. 3 (ed. G. Morel). CRC Press, Boca Raton, London, Tokyo, 35–56.

Pottu-Boumendil, J. (1989). *Microscopie Electronique: Principes et méthodes de preparation.* Les Editions INSERM, Paris.

Chapter 8
Comparisons of Techniques

1 Introduction

Given the different nature of the artifacts and drawbacks induced by mechanical, chemical, or ionic techniques, or even those involving changes in physical state, it is important to combine several techniques in order to confirm the intrinsic structure of a given material. The combination of techniques can vary, depending on the different properties of materials, their physical or chemical state, and their organization.

This chapter presents comparisons of at least two of the most commonly used preparation techniques for different types of materials in materials sciences and in biology.

2 Examples Using Fine Particle Materials

2.1 Comparison of Mechanical Preparations and Replicas

Extractive replica technique and crushing
Fine particle materials: catalyst particles

2.1.1 Crushing Technique ("Techniques" Chapter 4, Section 1) and Extractive Replica Technique ("Techniques" Chapter 5, Section 3)

Fine particle material: 5% platinum catalyst particles on cerium oxide (CeO_2), treated with H_2 at 973 K.

Comparison discussion: The crushing technique enables the conservation of the substrate and investigating the interactions between it and the material (catalyst particles or other). In this case, Figs. 8.1 and 8.2 clearly show that platinum particles

J. Ayache et al., *Sample Preparation Handbook for Transmission Electron Microscopy*,
DOI 10.1007/978-0-387-98182-6_8,

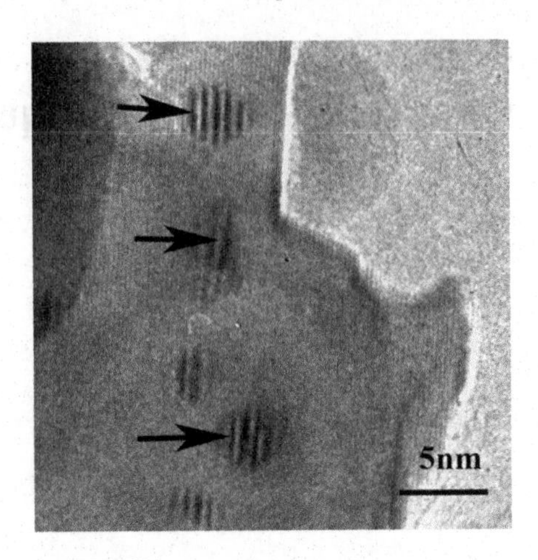

Fig. 8.1 The thin slice is obtained using the crushing technique and therefore, the substrate is maintained. The *arrows* indicate the platinum particles superimposed on the substrate, cerium oxide. The platinum particles are epitaxied with the substrate (moiré fringes). (*M. Abid, LMPC-ECPM-ULP Strasbourg*)

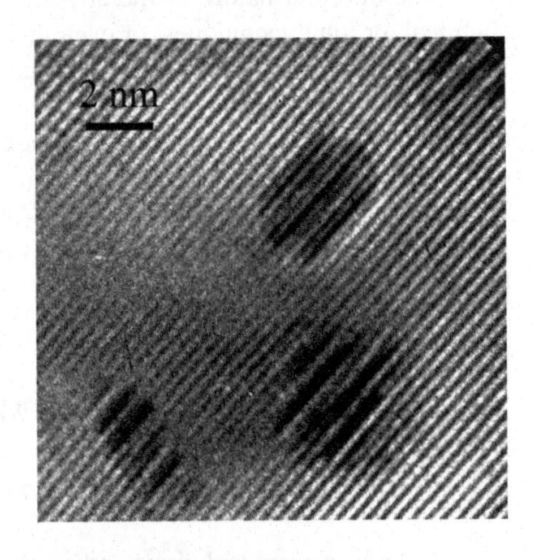

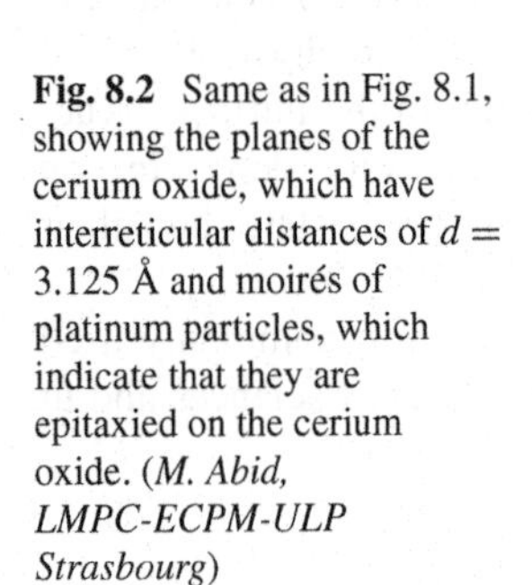

Fig. 8.2 Same as in Fig. 8.1, showing the planes of the cerium oxide, which have interreticular distances of $d =$ 3.125 Å and moirés of platinum particles, which indicate that they are epitaxied on the cerium oxide. (*M. Abid, LMPC-ECPM-ULP Strasbourg*)

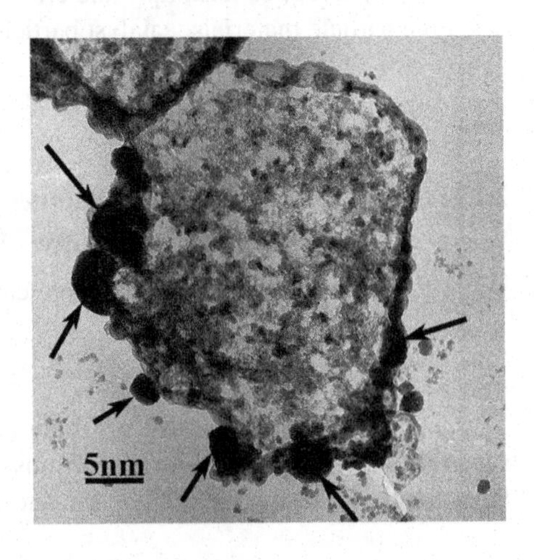

Fig. 8.3 The material is prepared using the extractive replica technique; the substrate is removed by means of chemical etching, only the shadowing of the cerium oxide is kept. The platinum particles are indicated by *arrows*. The particle–substrate interaction is lost and therefore, observation is more difficult. (*M. Abid, LMPC-ECPM-ULP Strasbourg*)

are epitaxied on the substrate. With the extractive replica technique, there is no way to investigate this interactivity since the substrate has to be dissolved (Fig. 8.3).

2.1.2 Crushing Technique ("Techniques" Chapter 4, Section 1) and Extractive Replica Technique ("Techniques" Chapter 5, Section 3)

Fine particle material: 0.2% platinum catalyst particles on aluminum oxide (Al_2O_3), treated with H_2 at 973 K

Comparison discussion: the crushing technique enables keeping the substrate together with the material. The platinum particle appears in black (high atomic number) and allows measurements and characterizations in nanodiffraction mode. The images of platinum in HRTEM do not have good resolution because the thickness of the platinum/substrate is too high. With the extractive replica technique, the substrate is dissolved and HRTEM observations are both possible and generally of good quality. Measurements of platinum particle size as well as characterizations in nanodiffraction mode can also be made. In this case, good resolution can easily be obtained in HRTEM images, since the substrate is removed (Figs. 8.4 and 8.5).

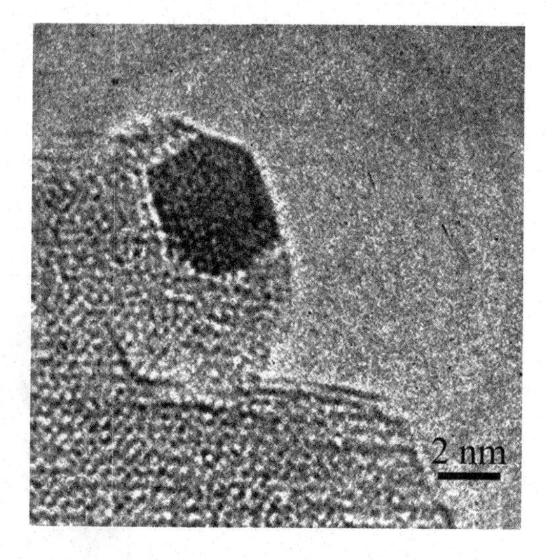

Fig. 8.4 Pt catalyst on aluminum oxide Pt/Al_2O_3. Thin slice obtained using the crushing technique. (*R. Touroude, LMPC-ECPM-ULP Strasbourg*)

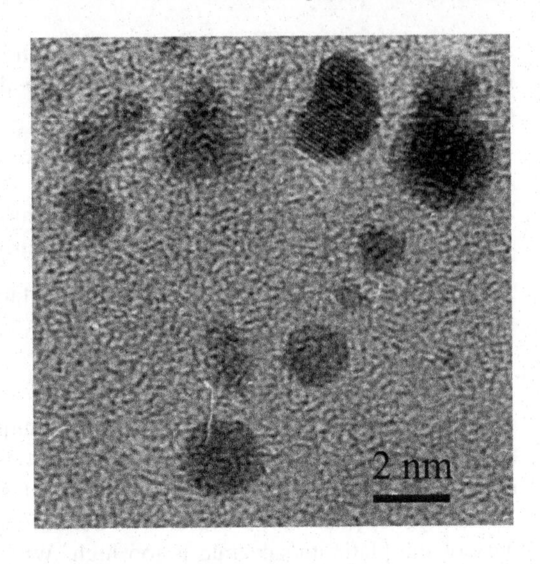

Fig. 8.5 Identical material, prepared by the extractive replica technique. (*R. Touroude, LMPC-ECPM-ULP Strasbourg*)

2.2 Comparison of "Negative-Staining" Contrast and Freeze-Fracture Techniques

2.2.1 Negative-Staining ("Techniques" Chapter 7, Section 2) and Freeze-Fracture Techniques ("Techniques" Chapter 5, Section 4)

Fine particle material: multilayer liposomes

Comparison discussion: The liposomes are spheres made from phospholipids. Here we have liposomes with multiple layers. Negative staining shows these different layers, but liposomes are often coalescent and fuse together (Fig. 8.6). Freeze fracturing allows us to see them separated from one another, as in the liquid suspension. It is easy to measure their size and evaluate their dispersion. Liposomes

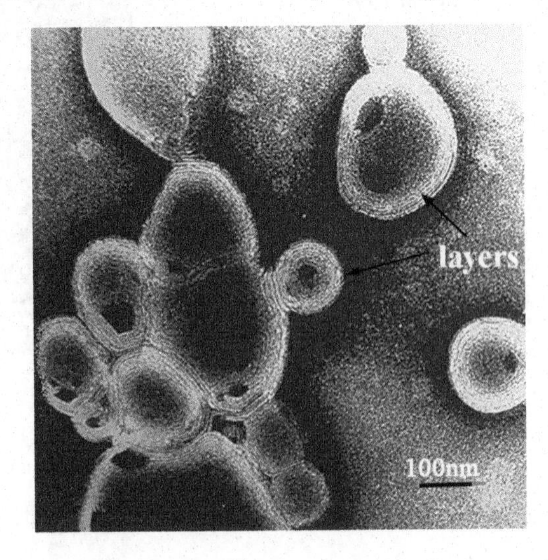

Fig. 8.6 The suspension of liposomes is treated using phosphotungstic acid. The liposomes are viewed using transparency; the layers are *white* on a *dark* background. The contrastant also penetrates inside the liposomes. (*J. Boumendil, UCB Lyon 1*)

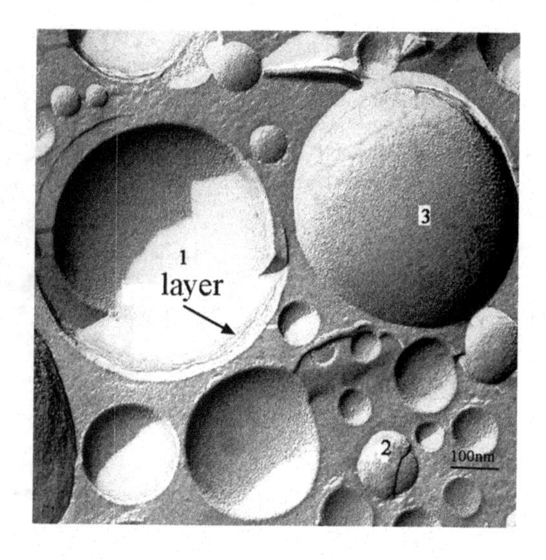

Fig. 8.7 The suspension of liposomes was frozen in nitrogen, and then freeze fractured. Some liposomes break and open completely (1), others are half-broken, between two layers (2), and others are intact (3), as viewed from above. (*J. Boumendil, UCB Lyon 1*)

fracture, making it possible to highlight the multiple layers (Fig. 8.7). But most often, the very small and very large liposomes do not fracture. Either their imprint (1) or their surface is seen (3).

2.3 Comparison of "Negative-Staining" and Decoration-Shadowing Contrast Techniques

2.3.1 Negative-Staining Contrast ("Techniques" Chapter 7, Section 2) and Decoration-Shadowing ("Techniques" Chapter 7, Section 1) Techniques

Fine particle material: spread-out chromatin

Comparison discussion: Figs. 8.8, 8.9, and 8.10 show that it is possible to use the negative-staining and shadowing techniques on comparable samples for certain biological materials. Figure 8.8 shows chromatin fibers with the classic 30-nm diameter. Figures 8.9 and 8.10 show a different state of the chromatin, when the conditions enable decondensing the nucleosomes, and the nucleosomes and internucleosomal DNA can then be seen.

The negative-staining technique, which is very well adapted to chromatin, is not as suitable for viewing chromosomes and DNA (although it would still be possible). Figures 8.9 and 8.10 provide two shadowing variants. Figure 8.9 shows a very classic shadowing technique using platinum and Fig. 8.10 shows that bidirectional shadowing using tungsten yields two thinner metallization layers (a few nanometers). This makes possible dark-field observation and therefore results in higher contrast and better resolution. These shadowing methods could also be used for chromatin, but it may undergo damage due to the placement of the sample under vacuum for metallization.

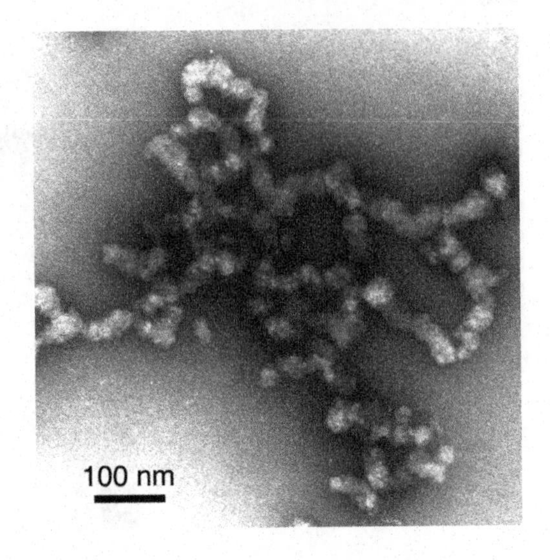

Fig. 8.8 Bright-field image of chromatin prepared using negative staining with 2% uranyl acetate in water. The protein complex is not unwound. (*E. Delain, CNRS-UMR8126-IGR, Villejuif*)

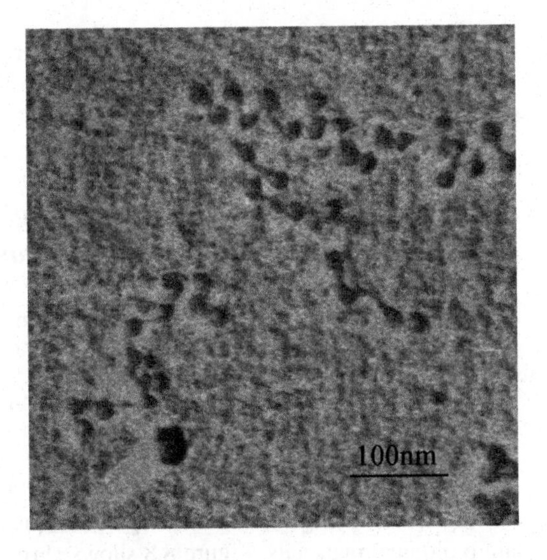

Fig. 8.9 Bright-field image of chromatin prepared using unidirectional shadowing with platinum. (*E. Delain, CNRS-UMR8126-IGR, Villejuif*)

2.3.2 Negative-Staining ("Techniques" Chapter 7, Section 2) and Decoration-Shadowing ("Techniques" Chapter 7, Section 1) Contrast Techniques

Fine particle material: *Escherichia coli* bacteria

Comparison discussion: Negative staining is used to quickly obtain information on the type of bacteria and flagella, but does not identify the pilis. Furthermore, the

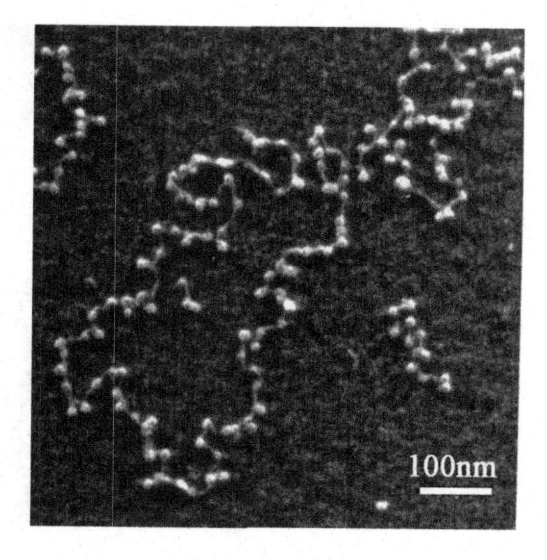

Fig. 8.10 Dark-field image of chromatin prepared using bidirectional shadowing with tungsten. (*E. Delain, CNRS-UMR8126-IGR, Villejuif*)

central body of the bacteria appears dehydrated and flattened (Fig. 8.11). Rotary shadowing shows the pili and flagella, but no information is available on the body of the bacteria; it would be necessary to make a replica in order to have this kind of data (Fig. 8.12).

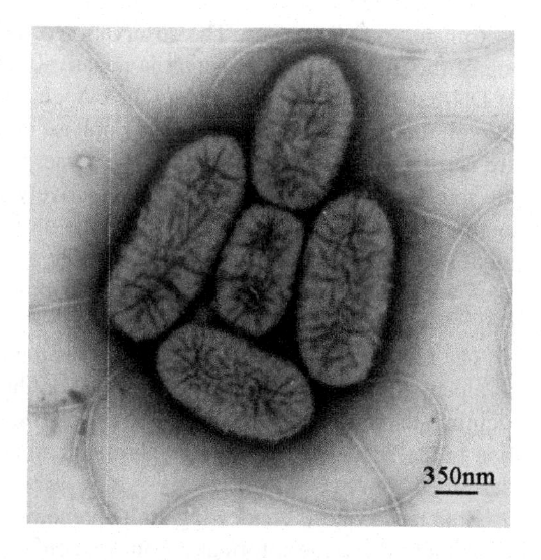

Fig. 8.11 Negative-staining, bright-field observation. (*A. Ryter, Institut Pasteur Paris*)

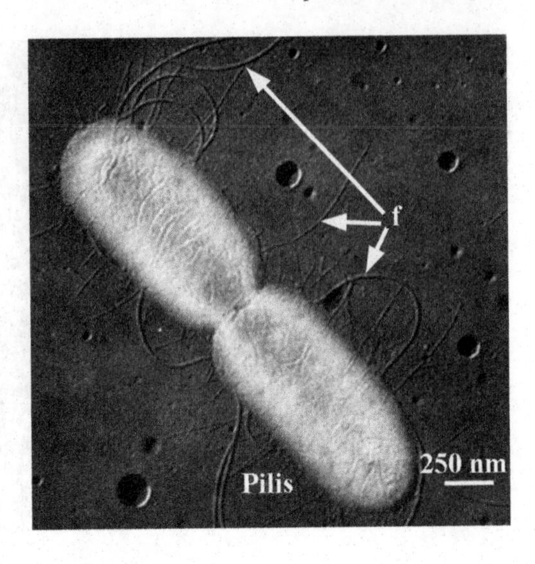

Fig. 8.12 Metallic rotary shadowing under a unidirectional angle of 45°, observed in bright-field and inverted image mode (f = flagella). (*A. Ryter, Institut Pasteur Paris*)

2.4 Comparison of "Positive-Staining" and Decoration-Shadowing Contrast Techniques

2.4.1 Positive-Staining ("Techniques" Chapter 7, Section 2) and Decoration-Shadowing ("Techniques" Chapter 7, Section 1) Contrast Techniques

Fine particle material: relaxed circular DNA molecules (pBR 322 plasmid, 4,361 bp)

Comparison discussion: The observation of DNA molecules requires the spreading of the macromolecule, which is negatively charged. Historically, the first type of DNA spreading was done using the cytochrome *c* technique, which forms a thin film on the surface of an aqueous hypophase and complexes with the DNA. There are several variants of this method (see fine particle dispersion technique). Next, the spread DNA can be shadowed or positively stained in order to be viewed. The contrast depends on these methods and the observation in bright-field or dark-field mode. For dark-field imaging, we can make a tilted dark field, or use the electron energy loss, either by using elastically transmitted electrons or on the uranium peak (Figs. 8.13 and 8.14).

The second type of DNA spreading uses the technique of a preliminary coating of a pentylamine film (Dubochet method) on the carbon grid, followed by positive staining using a uranyl acetate rinse. Positive staining can highlight the conformation of the DNA. This type of preparation can also be metalized i.e., shadowing as described above (Fig. 8.14).

Both techniques, "shadowing after spreading of cytochrome *c* and positive staining," yield the same information on the conformation of the DNA, with contrasts and resolutions depending on the sharpness of the shadowing and on the observation mode.

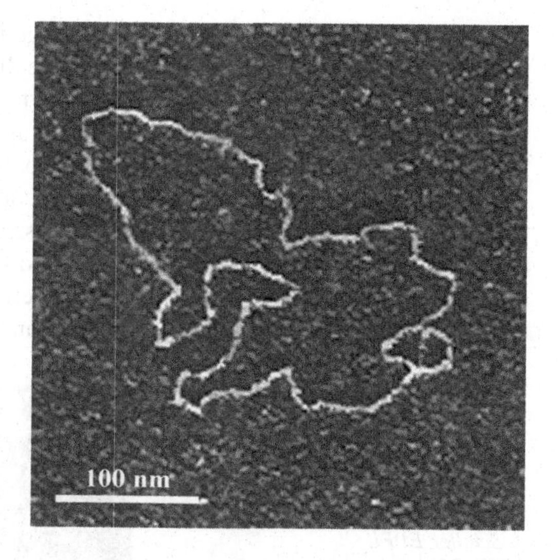

Fig. 8.13 Dark-field image of DNA after positive staining, followed by the decoration-shadowing technique using platinum/tungsten. (*E. Le Cam and E. Delain, CNRS-UMR8126-IGR, Villejuif*)

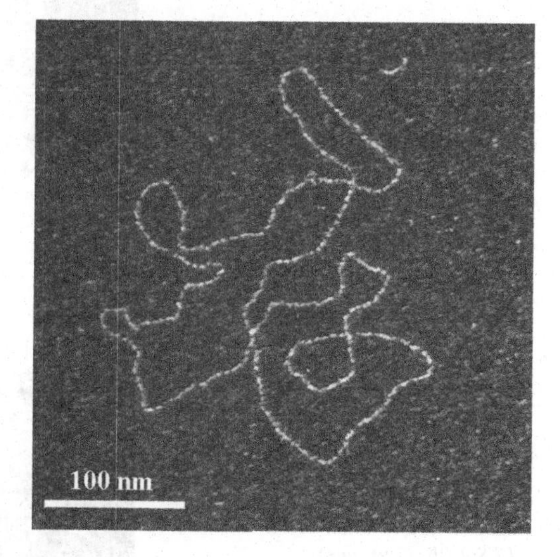

Fig. 8.14 Dark-field image of DNA after the contrast technique with uranyl acetate. (*E. Le Cam and E. Delain, CNRS-UMR8126-IGR, Villejuif*)

3 Examples Using Bulk or Multilayer Materials

3.1 Comparison Between Different Mechanical Preparations

Wedge Cleavage Technique ("Techniques" Chapter 4, Section 2) and Tripod Polishing Technique ("Techniques" Chapter 5, Section 3)

Mixed–composite multilayer material: Cross section of a sample of gold particles in a layer of silicon dioxide (SiO_2) on a silicon substrate

Comparison discussion: The wedge cleavage method does not make it possible to obtain a large observable surface area in the direction perpendicular to the edge of the wedge. Consequently, it is not possible to view a large number of particles. Furthermore, the thickness grows significantly and a superimposition of particles is then observed, excluding a statistical analysis of their size.

The tripod polishing technique, in a wedge configuration down to electron transparency, provides large thin observable areas, allowing for the investigation of dispersion and for statistics on particle size to be obtained. However, the wedge cleavage technique helps ensure that no material transformations have occurred during tripod mechanical thinning (Figs. 8.15 and 8.16).

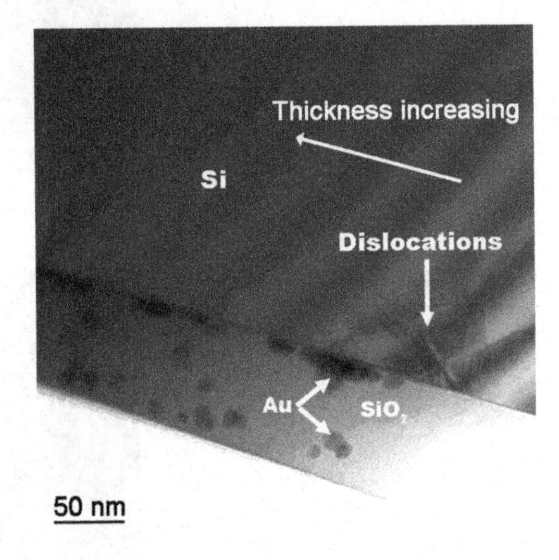

Fig. 8.15 Bright-field TEM image of the Au/SiO_2/Si sample, prepared using the cleaved wedge method (at 90° angle). (*S. de Chambrier, A. Schiller, EPFL – LESO-PB, Lausanne*)

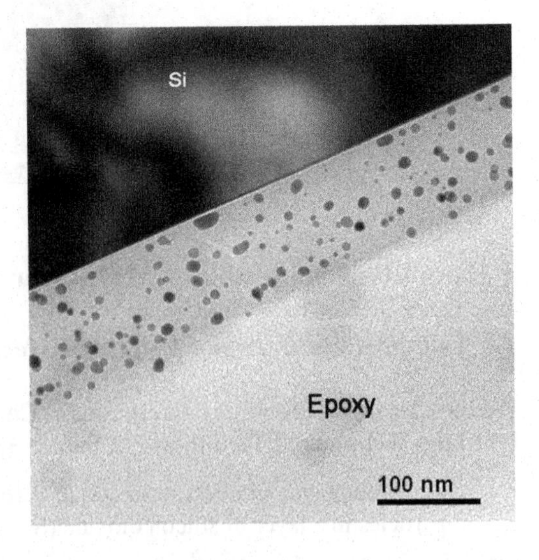

Fig. 8.16 Bright-field TEM image of the same sample prepared using the tripod polishing method, down to electron transparency. (*S. de Chambrier, A. Schiller, EPFL – LESO-PB, Lausanne*)

3.2 *Comparison Between Mechanical Preparations and Ionic Preparations*

3.2.1 Comparison of "Cleaved Wedge" and "Ionic Thinning" Techniques

Wedge Cleavage Technique ("Techniques" Chapter 4, Section 2) and Ion Milling Technique ("Techniques" Chapter 3, Section 5)

Multilayer material: Cross section of an AlGaAs/GaAs multilayer sample on a GaAs substrate

Comparison discussion: The cleaved wedge helps to highlight the variation in chemical composition (indicated by the arrows) near the AlGaAs quantum wells, visible by means of a change in thickness between the different fringes of equal thickness near the well. This variation is not visible on the image resulting from the ion milling preparation, due to the inhomogeneity of the sample thickness (Fig. 8.17).

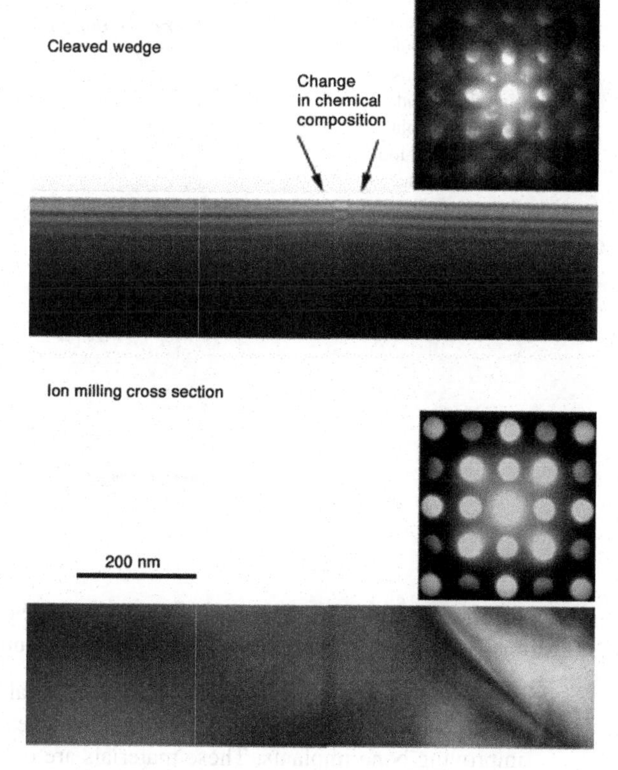

Fig. 8.17 Comparison of wedge cleavage (at 90°) and ion milling techniques for a sample of AlGaAs/GaAs on a GaAS substrate. Observation along the [100] direction. Bright-field TEM image. (*J-D. Ganière, EPFL – IPEQ, Lausanne, P.A. Buffat, EPFL-CIME, Lausanne*)

3.2.2 Comparison of "Tripod Polishing + Ions" and "FIB Thinning"

Tripod Polishing Technique + Ion Milling ("Techniques" Chapter 4, Section 3; Chapter 3, Section 5) and FIB Thinning Techniques ("Techniques" Chapter 3, Section 6)

Mixed–composite material in multilayer: Cross section of an oxide layer on a metallic substrate

Comparison discussion: Preparation using the tripod polishing method, followed by ion milling, was not sufficient enough to thin the metallic substrate in order to observe the oxide–substrate interface. Furthermore, the interface is not clearly delimited due to the slice being too thick. Lastly, there are redeposition particles (visible in the epoxy used for preparing the sandwich) of the material pulverized during ion bombardment, which will impede observation and chemical analysis. It cannot be known with certainty if the porosities (1) present in the oxide layer are inherent to the material or if they were caused by ion bombardment.

Preparation using the FIB technique is used to obtain a thinner slice than that obtained using ion bombardment. The relatively constant thickness of the lamella enables the investigation of the oxide–substrate interface and confirms the presence of porosities in the oxide layer (Figs. 8.18 and 8.19).

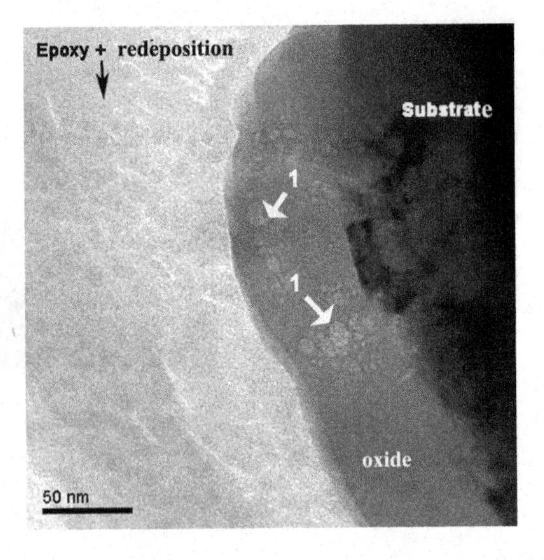

Fig. 8.18 Bright-field TEM image of an oxide/metal sample prepared in cross section using the tripod polishing technique, plane polished, followed by ion milling at low incidence angle, and low acceleration voltage (2 kV). (*M. Cantoni, EPFL – CIME, Lausanne*)

FIB Thinning Technique ("Techniques" Chapter 3, Section 6) and Tripod Polishing Technique ("Techniques" Chapter 4, Section 3)

Mixed–composite multilayer material (biomaterials): Hydroxyapatite (HA), doped with manganese and carbonate, is investigated in the context of improving bone implants. These materials are obtained by deposition using

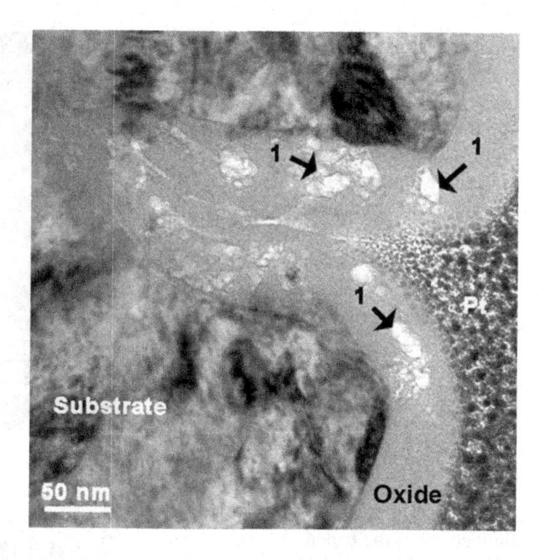

Fig. 8.19 Bright-field TEM image of the same sample prepared using the FIB technique and the "H Bar" method. Platinum has been deposited on the surface of the oxide to protect it during the thinning process. (*M. Cantoni, EPFL – CIME, Lausanne*)

laser ablation on titanium: Hydroxyapatite (HA) on titanium TiN/Ti or hydroxyapatite (HA) on TiN/Si

Comparison discussion: These materials present extreme physical–chemical properties and, consequently, are very difficult to prepare in thin slices for the TEM (Figs. 8.20, 8.21, and 8.22).

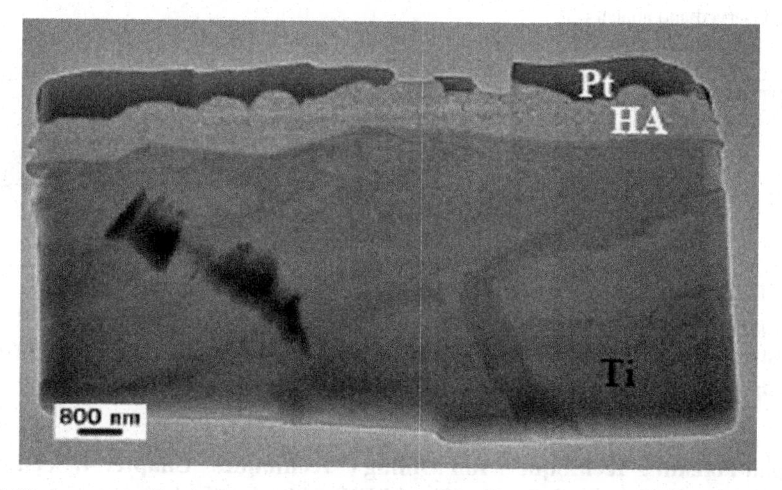

Fig. 8.20 Low-magnification image of the thin slice obtained using FIB. The HA/TiN/Ti material is thinned using the FIB technique. The hydroxyapatite layer is protected by a film of platinum which is evaporated in the FIB thinner, before being cut out by the ion beam, in order to protect it. The surface, however, is not very regular, and the Pt film has been removed in two areas. (*M. Iliescu, J. Werckmann IPCMS Strasbourg*)

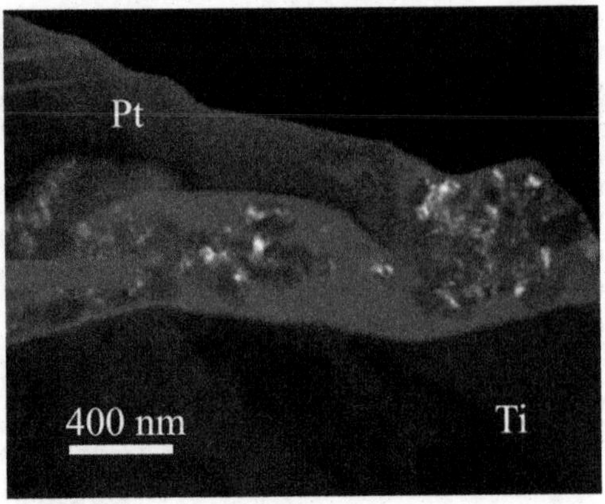

Fig. 8.21 Dark-field, higher-magnification image of the same material, showing that the hydroxyapatite is a mixture of amorphous and crystalline material. The FIB technique did not enable producing slices thin enough for HRTEM observations. (*M. Iliescu, J. Werckmann IPCMS Strasbourg*)

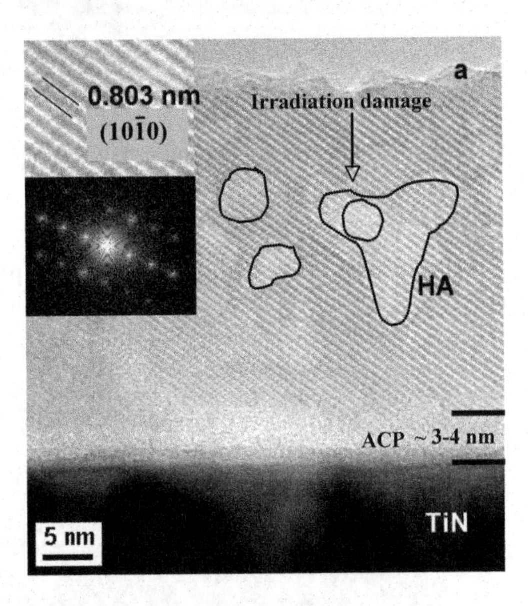

Fig. 8.22 The material is thinned using the tripod polishing technique, and a cross section of the HA/TiN/Si material is obtained (the Si substrate is not visible here). Only a fine layer of HA in contact with TiN (which is opaque) is kept, a large part of the material was torn away during the mechanical polishing. However, the surface (**a**) is well preserved and appears flat, and despite the damage from irradiation, observation in HRTEM is possible. (*M. Iliescu, J. Werckmann IPCMS Strasbourg*)

FIB can thin down very brittle materials to obtain large observable areas, but with poor resolution. Conversely, the tripod polishing technique enables HRTEM investigations, but only on very small areas, with the risk of material losses through tearing. Both techniques are complementary.

Tripod Polishing Technique + Ion Milling ("Techniques" Chapter 4, Section 3; "Techniques" Chapter 3, Section 5) and FIB Thinning Techniques ("Techniques" Chapter 3, Section 6)

Bulk material: Cross section of a metallic superconductor with Nb_3Sn filaments in a bronze matrix

Comparison discussion: The tripod polishing technique followed by a final thinning by ion milling results in a preferential etching: only the edges of the filaments are thin and the bronze matrix is milled preferentially, until its total disappearance in some areas (1).

The FIB technique makes possible a homogenous thinning of the filaments (the center of which is composed of Nb) and of the bronze matrix. A large number of filaments are visible (Figs. 8.23 and 8.24).

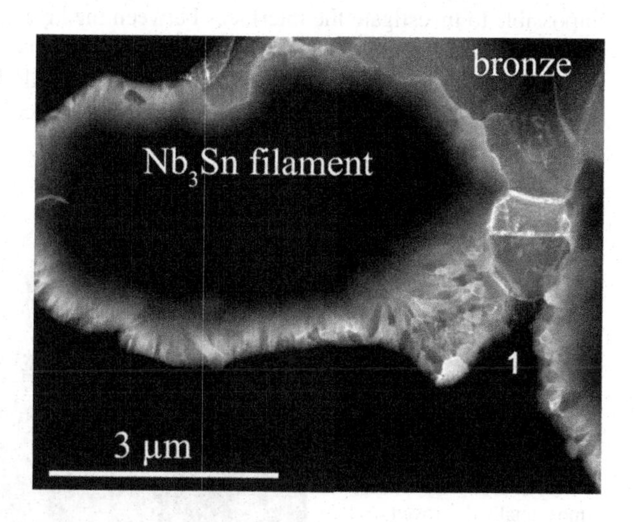

Fig. 8.23 Dark-field STEM picture. Preparation using the wedge tripod polishing method followed by ion milling under low incidence angle (5°). (*M. Cantoni, EPFL – CIME, Lausanne*)

Fig. 8.24 Dark-field TEM image of the same sample prepared using the FIB technique ("H Bar" method). (*M. Cantoni, EPFL – CIME, Lausanne*)

3.2.3 Comparison of "Ultramicrotomy" and "Ion Milling" Techniques

Ion Milling Technique ("Techniques" Chapter 3, Section 5) and Ultramicrotomy Technique ("Techniques" Chapter 4, Section 4)

Bulk material: cross section of a mica sample (mineral)

Comparison discussion: The mechanical polishing technique followed by the ion milling technique results in the almost total destruction of the mica layers. It is impossible to investigate the interfaces between the layers.

The ultramicrotomy technique yields excellent results: the layers are clearly visible, without alteration; therefore, the investigation of the interfaces between these layers is possible, as is the chemical analysis of the compounds (Figs. 8.25 and 8.26).

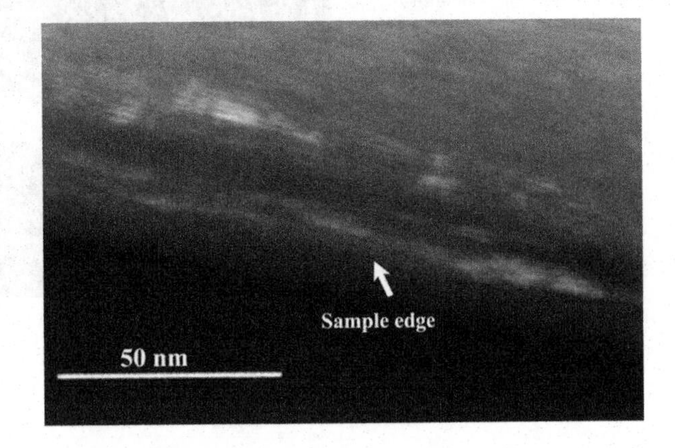

Fig. 8.25 Bright-field TEM image of a cross section of a mica sample thinned using mechanical polishing, followed by ion milling at 5 keV, two guns, angle of incidence 16°, and sectorial rotation (experimental conditions). The sample is thick. The layers of mica are almost totally destroyed. (*D. Laub, EPFL – CIME, Lausanne*)

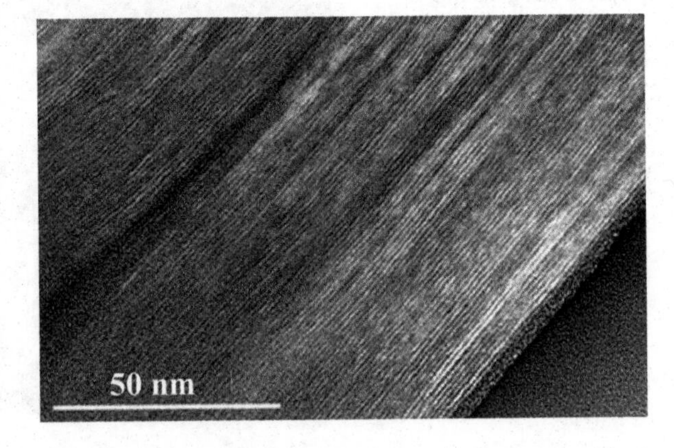

Fig. 8.26 Bright-field TEM image of the same sample prepared using the ultramicrotomy technique. Sample thickness: 70 nm. (*D. Laub, EPFL – CIME, Lausanne*)

3.2.4 Comparison of "Tripod Polishing," "Ion Milling," and "FIB Thinning" Techniques

Tripod Polishing Technique ("Techniques" Chapter 4, Section 3), Ion Milling Technique ("Techniques" Chapter 3, Section 5), and FIB Thinning Technique ("Techniques" Chapter 3, Section 6)

Mixed–composite multilayer material: Si/SiO_2/Ti/Pt/PZT/Pt (*sample E. Cattan Université de Valenciennes*)

Comparison discussion: The use of three techniques made it possible to show that none of them alone are sufficient. It is only the combination of techniques that makes it possible to validate the information on the physical properties of this material.

Tripod polishing makes it possible to highlight the mechanical behavior of the brittle metal/ceramic interfaces. We can see that the sample is still thick and does not allow the direct observation of the metal/ceramic interfaces, but only after a final ion thinning. However, observation of the microstructure on a large scale is possible.

The ionic thinning technique (cold dimpling + ions) makes it possible to obtain very thin samples in order to observe both the fine microstructural details and those of the porosity in the PZT film, using different microscopy modes.

The FIB helps maintain the interfaces despite the porosity, but results in a thick lamella that enables EDS analysis, but not EELS analysis (Figs. 8.27, 8.28, 8.29, 8.30, 8.31, and 8.32).

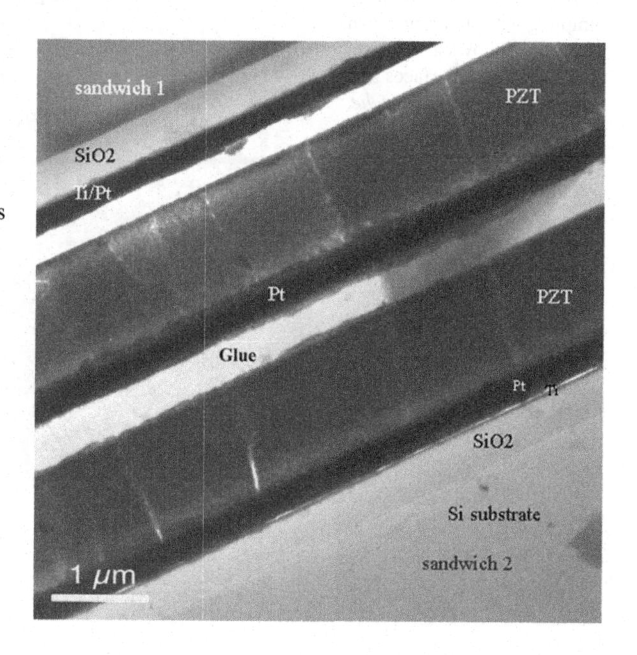

Fig. 8.27 Bright-field TEM image of a cross section of the multilayer material prepared using the tripod polishing technique, showing the entire sandwich at low magnification, which contains two whole interfaces on both sides of the adhesive. (*J. Ayache, S. Collin CNRS-CSNSM Orsay*)

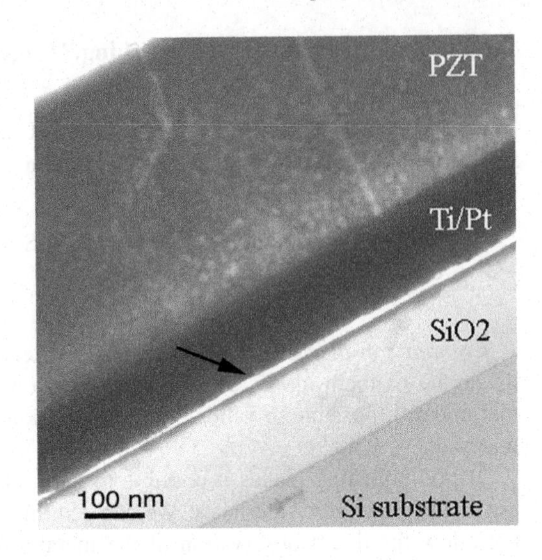

Fig. 8.28 Higher magnification of the image of the metal/ceramic interface from Fig. 8.27, showing that the Ti/Pt/SiO_2 interface is fractured. (*J. Ayache, S. Collin CNRS-CSNSM Orsay*)

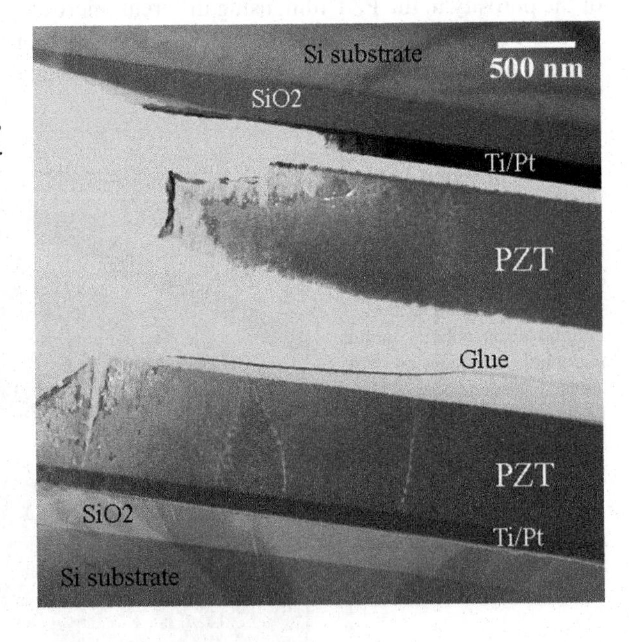

Fig. 8.29 Bright-field TEM image of a cross section of the multilayer material prepared using ionic thinning, dimpling with liquid nitrogen. This image shows a selective abrasion of the interfaces at low magnification (*J. Ayache, Cheng Y. Song NCEM, Berkeley*)

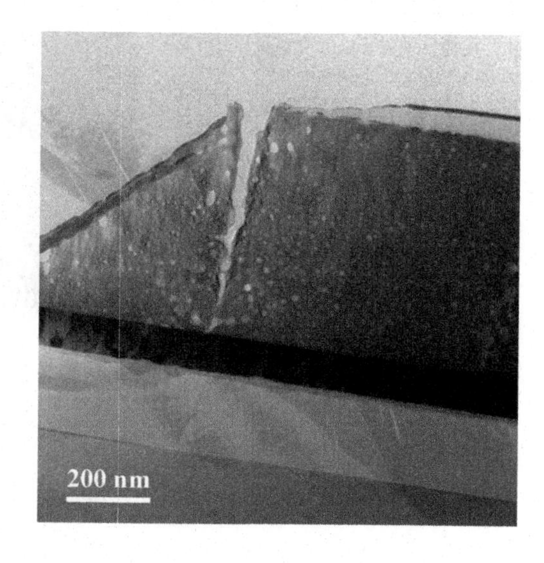

Fig. 8.30 Higher magnification of Fig. 8.29 showing thin areas observable in HRTEM in the PZT film, but hardly observable in the Ti/Pt/PZT interface (*J. Ayache, Cheng Y. Song NCEM, Berkeley*)

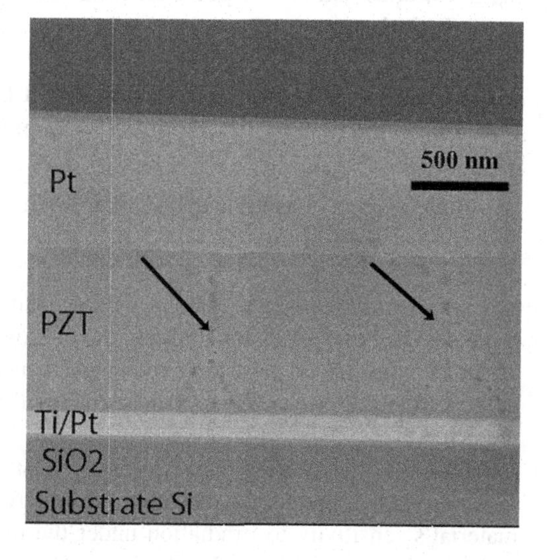

Fig. 8.31 SEM image in the FIB at the end of cross-sectional preparation on a pre-thinned slice. Porosities are visible in the PZT layer (*arrows*). (*V. Radmilovic, J. Ayache NCEM, Berkeley*)

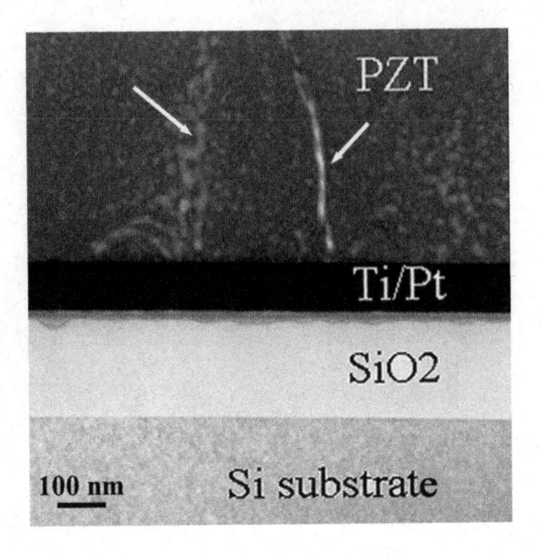

Fig. 8.32 TEM image of the same sample as shown in Fig. 8.31. This image shows thick interfaces over the entire surface as well as the porosities maintained in the PZT film observed in SEM. (*V. Radmilovic, J. Ayache NCEM, Berkeley*)

3.2.5 Comparison of "Tripod Polishing" and "Tripod Polishing + Ion Milling" Techniques

Tripod Polishing Technique ("Techniques" Chapter 4, Section 3) – Tripod Polishing + Ion Milling ("Techniques" Chapter 3, Section 5)

Bulk composite ceramic material: $YBa_2Cu_3O_7/Y_2BaCuO_5$

Comparison discussion: In both of these examples, the tripod polishing technique produces the best results for studying the microstructure of the interfaces.

In fact, the tripod polishing technique can be used to analyze all interfacial defects (dislocations, stacking faults, etc.) without irradiation damage and over a large area.

The ion bombardment technique never results in completely observable interfaces around a precipitate because of preferential etching and because of the material's sensitivity to irradiation under the ion beam, which promotes diffusion and selective loss of atoms. On this type of preparation, the electron beam provides annealing following diffusion and stabilizes the defects formed that are extrinsic to the starting microstructure.

In the case of tripod polishing technique, there is no annealing phenomenon under the electron beam, and the structure remains stable during observation (Figs. 8.33 and 8.34).

Fig. 8.33 Bright-field TEM image of an interface of a $YBa_2Cu_3O_7$ matrix/Y_2Ba–CuO_5 precipitate prepared by the tripod polishing technique. This image shows all of the interfaces surrounding the precipitate and the matrix twinning, ending in dislocations on the surface of the precipitate. (*J. Ayache, CNRS, CSNM, Orsay*)

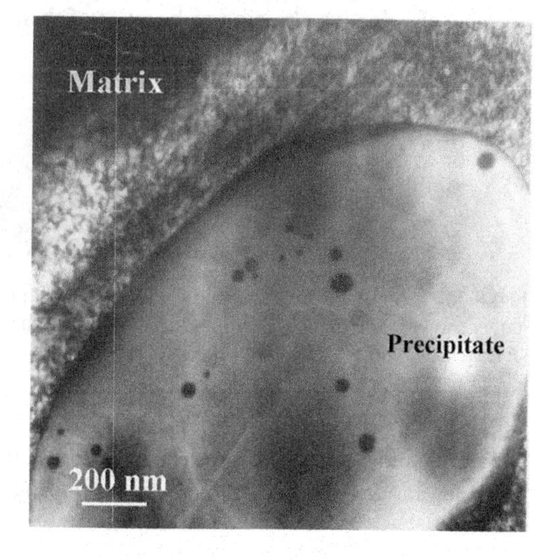

Fig. 8.34 Bright-field TEM image of a $YBa_2Cu_3O_7/Y_2BaCuO_5$ interface prepared using the ion milling technique, showing selective abrasion of the interface, and significant irradiation damage in the matrix and precipitate. (*J. Ayache, CNRS, CSNM, Orsay*)

3.2.6 Comparison of "Tripod Polishing" and "Ion Milling" Techniques

Tripod Polishing Technique ("Techniques" Chapter 4, Section 3) – Ion Milling Technique ("Techniques" Chapter 3, Section 5)

Multilayer composite ceramic material: $YBa_2Cu_3O_7/SrTiO_3$

Comparison discussion: Obviously, the tripod polishing technique produces the best results on the microstructure of the interfaces in the multilayer material (Fig. 8.35). In fact, this technique can be used to analyze all of the defects at the

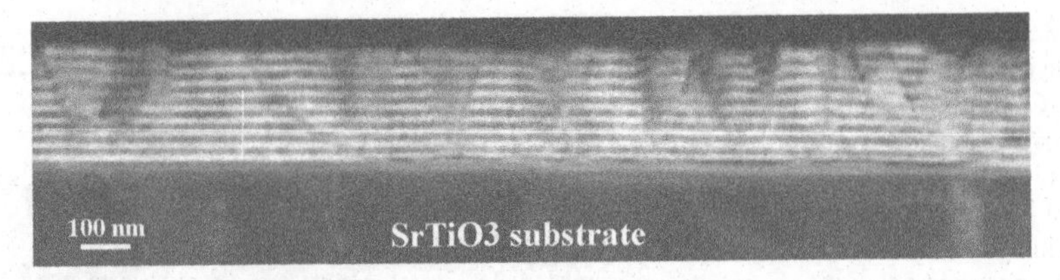

Fig. 8.35 Bright-field TEM image of a cross section prepared using the tripod polishing technique. This image shows all of the $YBa_2Cu_3O_7/SrTiO_3$ interfaces in a multilayer material created using laser ablation and shows regular growth defects. (*J. Ayache, CNRS, CSNM, Orsay*)

interfaces (dislocations, stacking faults, etc.) along the surface of the film and over a great distance along the multilayer material. It can also be used to obtain statistics on growth defects.

The ion milling technique results in heavy irradiation under the ion beam, which promotes the diffusion of atoms at the interfaces, as well as in the loss of crystal defects highlighted by the tripod polishing technique. Furthermore, the interfaces are no longer completely observable along the film/substrate interface, either because of selective abrasion or because of their amorphization (Fig. 8.36).

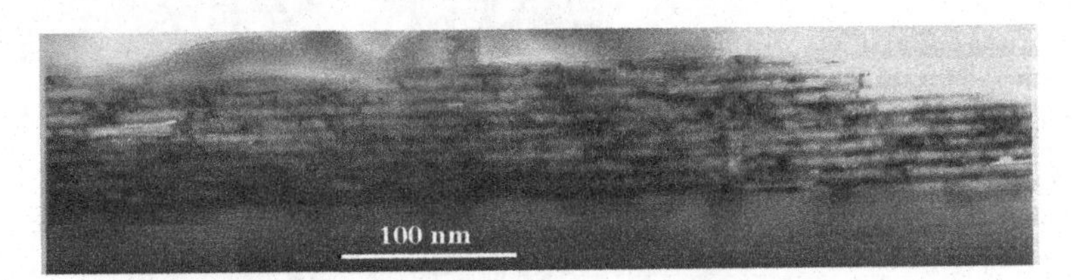

Fig. 8.36 Bright-field TEM image of a cross section prepared using the ion bombardment technique, showing the damage from ion irradiation and the loss of layers in the multilayer material. (*J. Ayache, CNRS, CSNM, Orsay*)

With the tripod polishing technique, there is no longer the phenomenon of annealing under the electron beam. In addition, the structure is stable, which enables the quantitative chemical analysis of the interfaces.

3.3 Comparison Between Mechanical Preparations and Electrolytic Preparations

Electrolytic Thinning Techniques ("Techniques" Chapter 3, Section 1), Tripod Polishing Technique + Ion Milling Technique ("Techniques" Chapter 4, Section 3; "Techniques" Chapter 3, Section 5), and Ultramicrotomy Technique ("Techniques" Chapter 4, Section 4)

Bulk metallic material: ZrNi alloy and ZrNi + CeO (Figs. 8.37, 8.38, 8.39, and 8.40)

Comparison discussion: In both of these examples, the tripod polishing technique combined with ion milling yields the best results on the intrinsic microstructure. However, although ultramicrotomy presents artifacts resulting in undulations and compression, this technique preserves the chemical information. Electrolytic thinning is harmful in this case (Figs. 8.37, 8.38, 8.39, and 8.40).

Fig. 8.37 Bright-field TEM image of a ZrNi planar view prepared using electrolytic thinning, showing the appearance of an unstable phase causing the formation of many networks of twins. (*Nathalie Michel, CNRS, LEMHE, Orsay*)

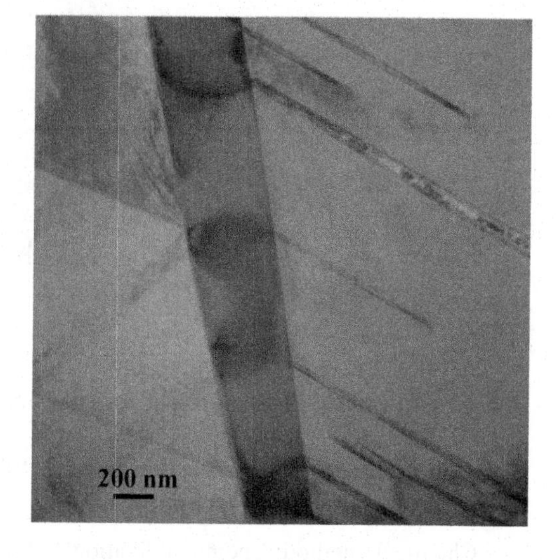

Fig. 8.38 Bright-field TEM image of a ZrNi planar view prepared using tripod polishing mechanical thinning followed by ionic thinning (PIPS), showing the intrinsic microstructure of the metallic alloy. (*Nathalie Michel, CNRS, LEMHE, Orsay*)

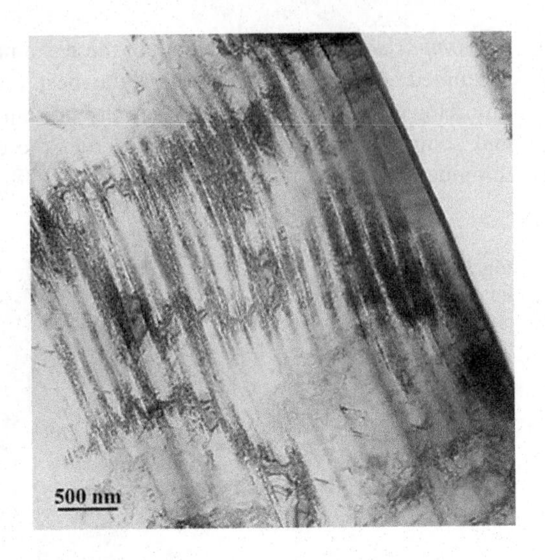

Fig. 8.39 Bright-field TEM image of a planar view of ZrNi + CeO prepared by tripod polishing + ion milling, showing an extended defect structure. (*Nathalie Michel, CNRS, LEMHE, Orsay*)

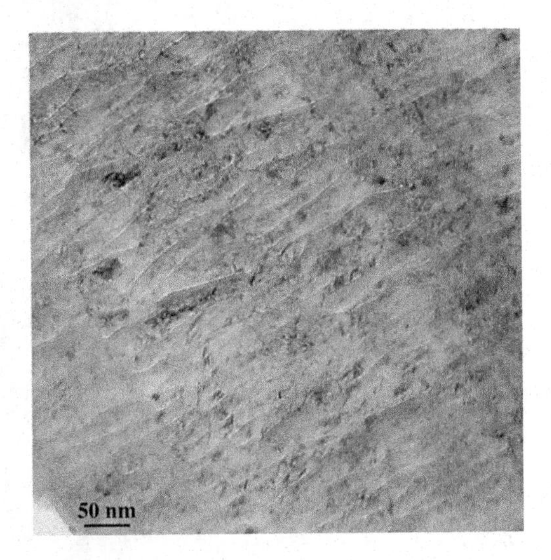

Fig. 8.40 Bright-field TEM image of a ZrNi + CeO planar view prepared using ultramicrotomy, showing undulations and compressions in the microstructure. (*Nathalie Michel, CNRS, LEMHE, Orsay*)

3.4 Comparison Between Techniques Specific to Biology

3.4.1 Comparison of "Chemical Methods," "Physical Methods," and "Freeze-Fracture" Techniques

Chemical method: chemical fixation ("Techniques" Chapter 2, Section 11) and substitution, infiltration, and embedding at room temperature ("Techniques" Chapter 2, Section 9)

Physical method: cryofixation ("Techniques" Chapter 2, Section 12) and substitution, infiltration, and embedding in cryogenic mode ("Techniques" Chapter 2, Section 10)
Ultramicrotomy ("Techniques" Chapter 4, Section 4) Positive-staining contrast ("Techniques" Chapter 7, Section 3)
Bulk material: liver cell (Figs. 8.41, 8.42, and 8.43)

Comparison discussion: The liver cell can be well identified using all three techniques. The membranes are different depending on the case. They are darkened by the presence of osmium in the chemical fixation and are very well marked in the nuclear membrane as well as the rough endoplasmic reticulum and mitochondria (Fig. 8.41). With cryofixation, the membranes are not as visible as such, but they are marked because they are highlighted by the surface glycoproteins, especially at the junction between two cells. Along the cell membranes (M1 and M2), a desmosome before the biliary canaliculus (c) can be distinguished. The nuclear membrane is not visible in the nucleus, and the placement of the nuclear pores is marked by the absence of heterochromatin (hC) in the nucleus at this level. The mitochondrial crests are dark by inverting the contrast, the membranes are bright, and the matrix is dark (Fig. 8.42).

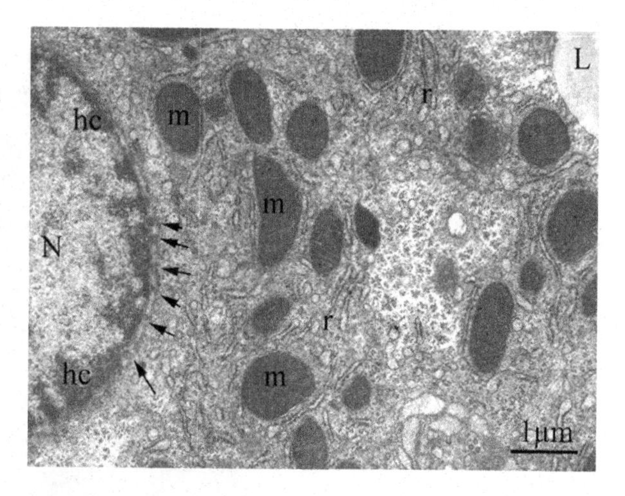

Fig. 8.41 Chemical fixation, substitution–impregnation, and embedding in epoxy resin at room temperature, ultramicrotomy and positive contrast (uranyl in alcohol solution + lead); hC = heterochromatin, L = lipidic phase, m = mitochondria, N = nucleus, r = rough endoplasmic reticulum, *black arrows* = nuclear pores. (*J. Boumendil UCB-Lyon 1*)

Freeze fracture penetrates the membrane in the double phospholipidic layer. Thus, in the nucleus, the external face (Fe) and the internal face (Fi) of the nuclear envelope can be seen, as well as the nuclear pores resembling buttons on the nucleus. They constitute a kind of canaliculi across the double nuclear membrane. This type of image helps to provide a good approach to the distribution of nuclear pores on a cell nucleus to take counts and to establish relative proportions (Fig. 8.43).

The chemical method is particularly used for investigating structure and for immunolabeling. The freeze-fracture technique is used for 3D viewing and the approach to intramembrane structures.

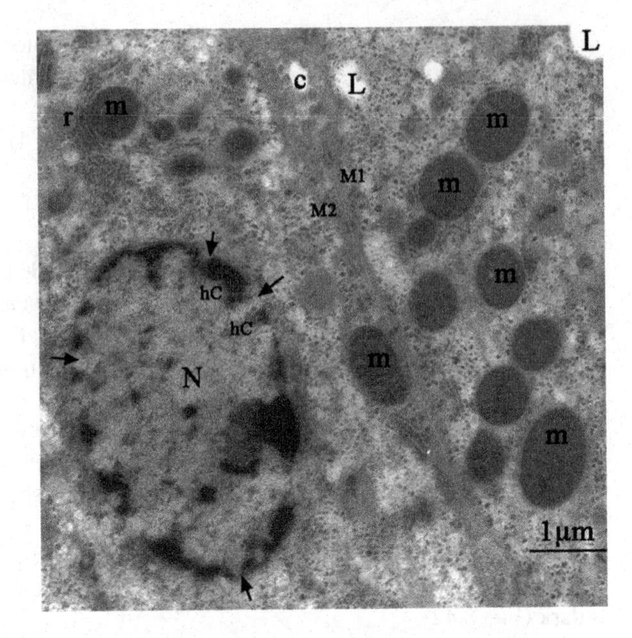

Fig. 8.42 Ultrarapid high-pressure cryofixation, cryosubstitution, and cryo-embedding in Lowicryl K4M, ultramicrotomy and positive-staining contrast (uranyl in aqueous solution + lead). (*J. Boumendil UCB-Lyon 1*)

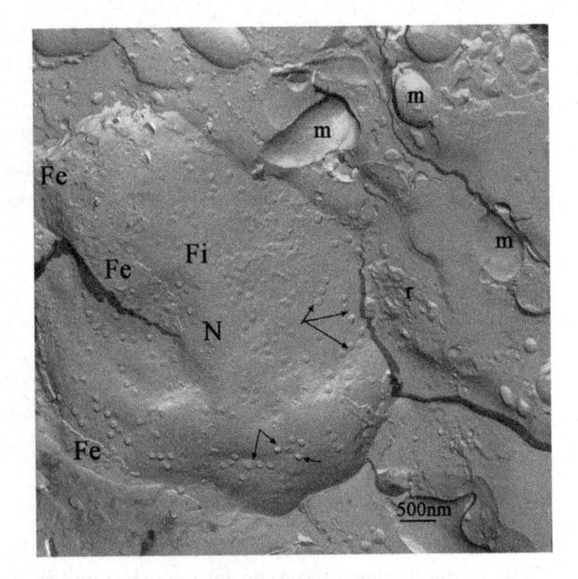

Fig. 8.43 Cryofracture at 123 K using the double cupel technique, platinum shadowing at 45°, and carbon replica. Fe = external face, Fi = internal face. (*A. Rivoire UCB-Lyon 1*)

3.4.2 Comparison of "Chemical Methods," "Physical Methods," and "Freeze-Fracture" Techniques

Chemical method: chemical fixation ("Techniques" Chapter 2, Section 11) and substitution, infiltration, and embedding at room temperature ("Techniques" Chapter 2, Section 9)

Physical method: chemical fixation ("Techniques" Chapter 2, Section 12) and substitution, infiltration, and embedding in cryogenic mode ("Techniques" Chapter 2, Section 10).
Ultramicrotomy ("Techniques" Chapter 4, Section 4) and positive-staining contrast ("Techniques" Chapter 7, Section 3)
Bulk material: rat duodenum

Comparison discussion: The duodenum cells are connected to one another by a succession of different types of junctions with different roles. A tight junction is found starting at the duodenal lumen bordered by microvilli. This is an impermeable junction that does not allow any elements from the alimentary bolus to pass between the cells. Then, a desmosomal junction, which is very resistant, is found. Lastly, a gap junction, which allows for rapid exchanges between two adjacent cells, is found.

The tight junction is recognized using chemical fixation and the classic preparation (Fig. 8.44) because the space between the two cell membranes is reduced. The desmosome is recognized because the intercellular space is enlarged and the plasmic membrane is reinforced on each side by tonofilaments; however, the gap junction is not recognized.

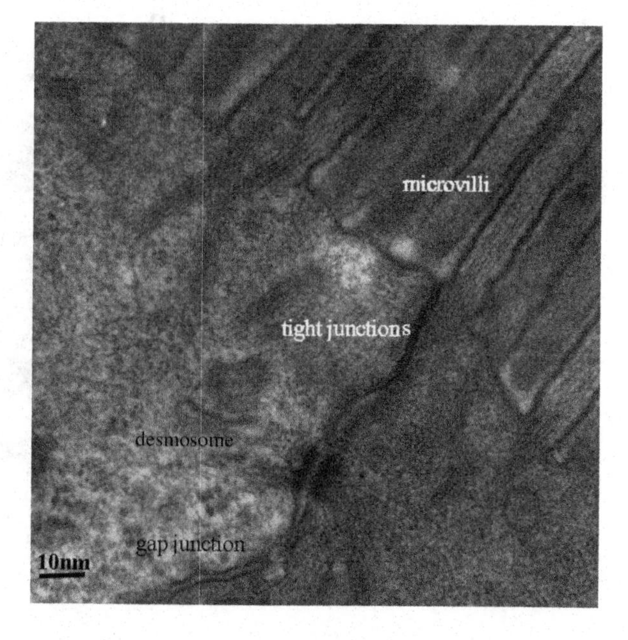

Fig. 8.44 Chemical fixation, substitution–infiltration–embedding in epoxy resin at room temperature, ultramicrotomy and positive-staining contrast (uranyl in alcohol solution + lead). The junctions between two adjacent cells can be seen. (*J. Boumendil UCB-Lyon 1*)

With ultrarapid cryofixation, the lack of osmium in the preparation does not make it possible to see the cell membranes (Fig. 8.45).

With freeze fracture, the inside of the cell membrane is visible and in particular, the transmembrane proteins can be seen in the form of small granules. At the level of the tight junction, the transmembrane proteins form a support network (Fig. 8.46b). In the area of the desmosome, there is no transmembrane protein; there is only an

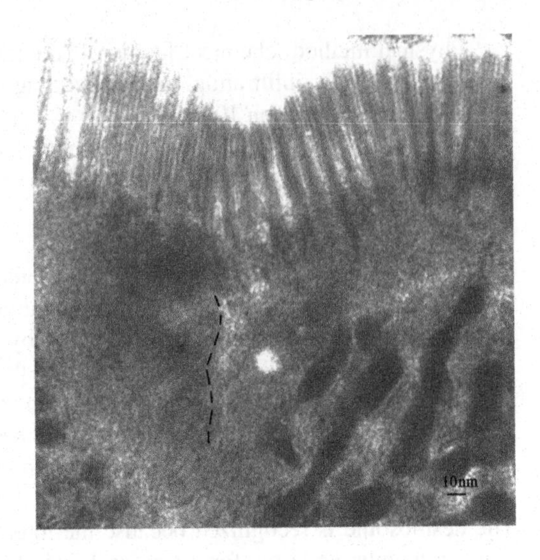

Fig. 8.45 Ultrarapid high-pressure cryofixation, cryosubstitution, and cryo-embedding in Lowicryl K4M, ultramicrotomy and positive contrast (uranyl in aqueous solution + lead). The cellular junction zone is marked by the *broken line*. (*J. Boumendil UCB-Lyon 1*)

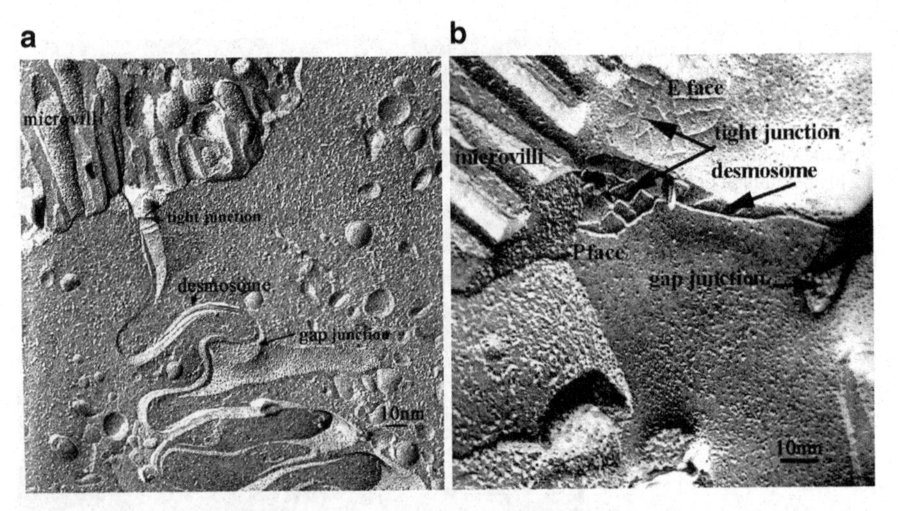

Fig. 8.46 (**a**) and (**b**) Freeze fracture at 123 K using the double cupel technique. Platinum shadowing at 45° and carbon replica. Cell junction between two enterocytes. (*A. Rivoire UCB-Lyon 1*)

intercellular space between the two membranes enabling it to be located. Lastly, at the level of the gap junction, there are a large number of transmembrane proteins (Fig. 8.46a) which organize into connections in order to form communicating canals from one cell to another. These canals ensure the rapid transport of chemical or electrical elements between the two cells. These properties were described by electrophysiologists before the corresponding structure had been seen.

The different techniques are necessary for viewing all of the structures of the membrane junctions and in order to take their functions into account.

3.4.3 Comparison of "Chemical Fixation," "Cryo-embedding," and "Immunolabeling" Techniques

Chemical method: chemical fixation ("Techniques" Chapter 2, Section 11) and substitution, infiltration, and embedding in cryogenic mode ("Techniques" Chapter 2, Section 10).
Ultramicrotomy ("Techniques" Chapter 4, Section 4) and positive-staining contrast ("Techniques" Chapter 7, Section 3) and Immunolabeling ("Techniques" Chapter 7, Section 4)
Bulk material: normal human dermis

Comparison discussion: Labeling is performed on ultrathin sections after "progressive low temperature" (PLT) embedding in Lowicryl K4M. At first, a chemical method is used, and then a cryo-method is used for the embedding. Hyaluronic acid is combined with a protein fragment, making it possible to make an antibody labeled with 10-nm gold particles. Hyaluronic acid is highly present in the dermis; it is found around the fibroblast, type 1 collagen fibers, and elastin, as shown in Fig. 8.48. It has an important role in maintaining skin tone, which is why it is frequently used in cosmetology. The advantage of the PLT method is that it can be used to properly carry out immunolabeling on extracellular proteins which are extracted by purely chemical methods. PLT method preserves the ultrastructure close to that of the purely chemical method (Fig. 8.47).

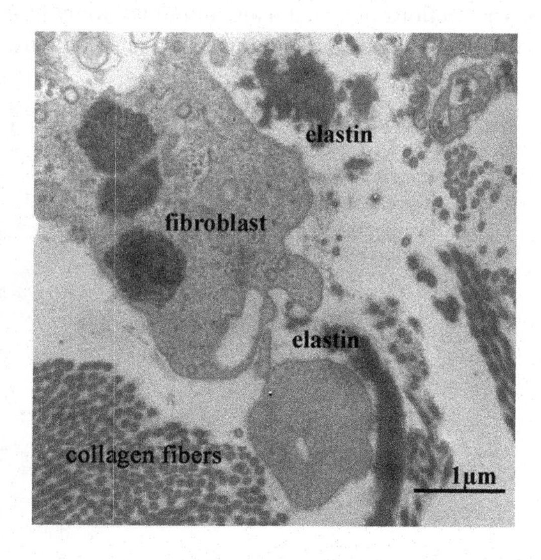

Fig. 8.47 TEM image of a portion of dermis with a cell called a fibroblast and the extracellular glycoproteins elastin and collagen prepared by ultramicrotomy using classic chemical fixation and embedding at room temperature. (*M. Haftek, Dermatologie UCB-Lyon 1*)

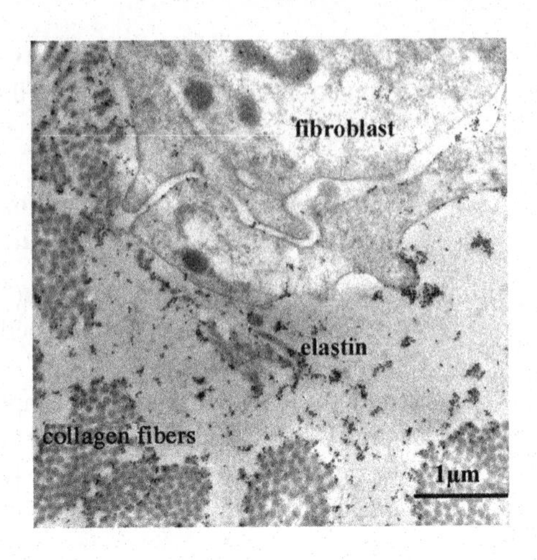

Fig. 8.48 TEM image of the same tissue with immunolocalization of hyaluronic acid-based proteins conducted on an ultrafine section prepared using chemical fixation, cryo-substitution and cryo-embedding and ultramicrotomy at room temperature. (*M. Haftek, Dermatologie UCB-Lyon 1*)

3.4.4 Comparison of "Immunolabeling," "Ultramicrotomy," and "Cryo-ultramicrotomy" Techniques

Techniques: Ultramicrotomy ("Techniques" Chapter 4, Section 4), Cryo-ultramicrotomy ("Techniques" Chapter 4, Section 5), and Immunolabeling ("Techniques" Chapter 6, Section 4)

Bulk material: normal human dermis

Comparison discussion: Both figures show spiny junctions located between two keratinocyte epidermal cells. The same immunolabeling is conducted on both types of preparation. Here, an inter-keratinocyte epidermal proteoglycan is labeled with an antibody coupled to 10-nm gold particles. In the first case, the tissue has been chemically fixated and embedded at low temperature (PLT method) in Lowicryl K4M. The sections obtained using ultramicrotomy have been immunolabeled (Fig. 8.49). In the second case, the tissue was fixated only using paraformaldehyde before being

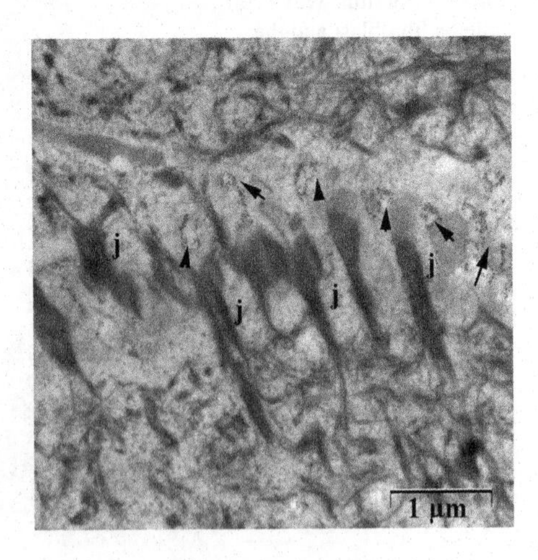

Fig. 8.49 Immuno-localization of an epidermal proteoglycan in the junction zone (j) of two keratinocytes. Immunolabeling on the ultrafine section after PLT embedding and ultramicrotomy. (*M. Haftek, Dermatologie UCB-Lyon 1*)

cryo-protected and frozen in liquid propane. The sections were made using cryo-ultramicrotomy and then they were immunolabeled (Fig. 8.50). Both methods result in the immunolocalization of the protein in the junction area. In the first method, the protein has been moved and regrouped by zone, whereas in the second method, it has remained dispersed, which is explained by the absence of dehydration in protocol 2. Conversely, in the first method, the structure is easier to interpret and a comparison helps to localize the area of interest in the second method.

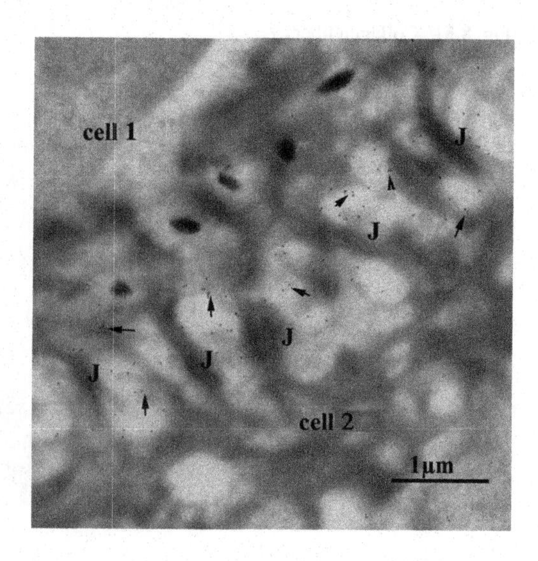

Fig. 8.50 Immuno-localization of the same junction zone (J). Immunolabeling on frozen tissue and an ultrafine section obtained using cryo-ultramicrotomy. (*M. Haftek, Dermatologie UCB-Lyon 1*)

3.5 Comparison Between All Techniques That Can Be Used in Biology on One Example: Collagen

Comparison of All Techniques Used in Biology on One Example: Collagen

Collagen fibers are components of support tissues of an animal organism (bone, skin, cartilage, connective tissue, etc.). So far, 29 or 30 different types of collagen can be distinguished.

3.5.1 Comparison of "Negative-Staining Contrast" and "Immunolabeling techiques"

"Negative-staining" contrast technique ("Techniques" Chapter 7, Section 2) and immunolabeling ("Techniques" Chapter 7, Section 4)

Fine particle material: isolated collagen fibers (Figs. 8.51 and 8.52).

Comparison discussion: Cartilage is composed of a mixture of collagen 2 and collagen 11. Collagen 2 is in the form of large periodic fibers similar to those of collagen 1. When collagen 2 fibers are attacked chemically, small fibrils appear that can be seen with a negative staining (Fig. 8.51). The immunological labeling made on a specific antibody of collagen 11 proves that these small fibrils are collagen 11 fibrils (Fig. 8.52).

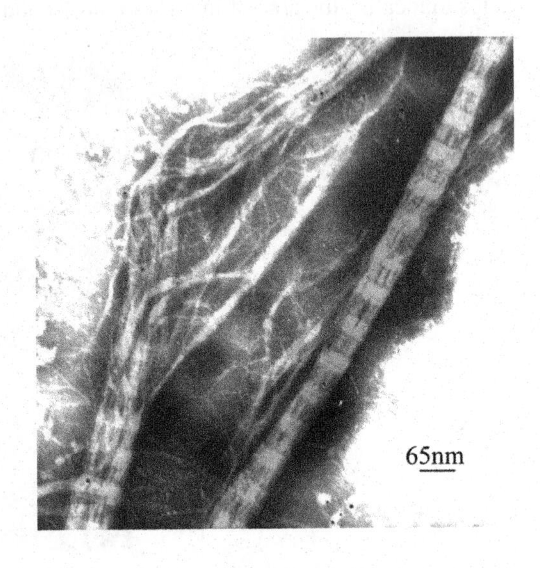

Fig. 8.51 Collagen fibers 2, partially dissociated in order to allow the appearance of small collagen fibrils 11, after "negative-staining" contrast. (*B. Burdin, CTμ UCB-Lyon*)

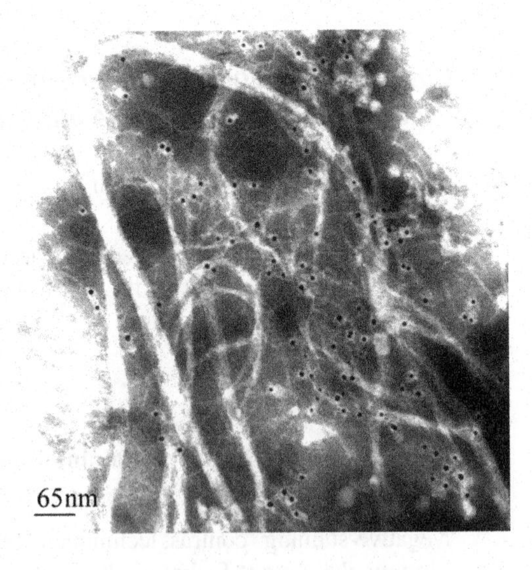

Fig. 8.52 Collagen 11 is labeled using a specific antibody coupled to gold particles (in *black* in the photo) with "negative-staining" contrast. (*B. Burdin, CTμ UCB-Lyon*)

3.5.2 Comparison of "Negative-Staining" and "Decoration-Shadowing" Contrast, and "Freeze-Fracture" Techniques

Negative-Staining Technique ("Techniques" Chapter 7, Section 2), Decoration Shadowing ("Techniques" Chapter 7, Section 1), and Freeze Fracture ("Techniques" Chapter 5, Section 4)

Fine particle material: collagen fibers in a tissue or isolated fibers (Figs. 8.53, 8.54, and 8.55)

Comparison discussion: Here we have type 1 collagen fibers isolated from tissue. The negative-staining technique makes it possible to see the fiber and its periodicity (Fig.8.53). Rotary shadowing also reveals the microfilaments or globular proteins that can be associated with the fiber (Fig. 8.54). This technique is especially interesting when working with molecules of small dimensions (other types of collagen, DNA, RNA, and other fibrillary proteins). The freeze-fracture technique yields an image "in relief," with clearer contours that favor measurements. It also makes it possible to view the globular proteins that might be associated (Fig 8.55). It is possible to perform immunolabeling on the freeze fracture.

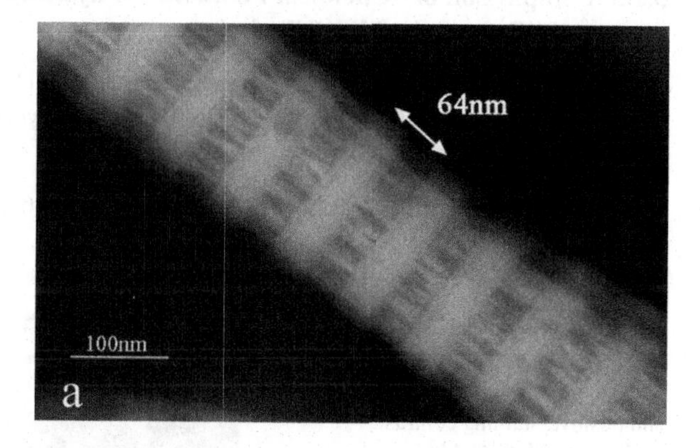

Fig. 8.53 Reconstituted collagen fiber in negative staining. (*A. Rivoire EZUS-UCB Lyon 1*)

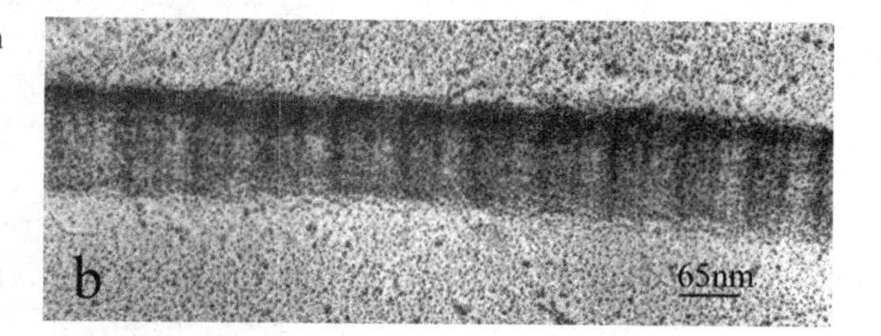

Fig. 8.54 Collagen fiber with rotary shadowing. (*B. Burdin CTμ UCB-Lyon 1*)

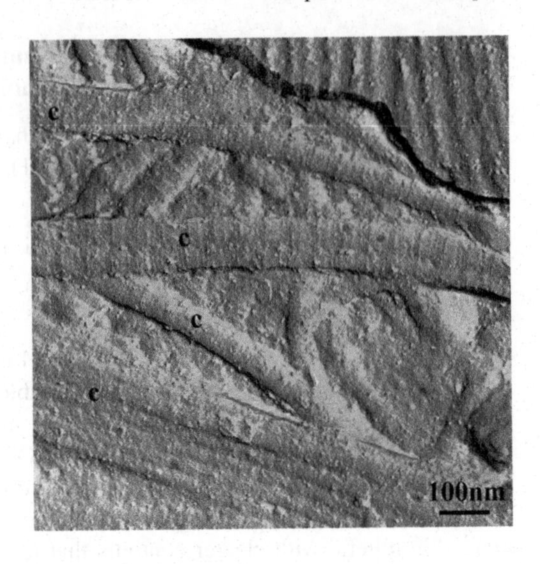

Fig. 8.55 Collagen fibers (c) after freeze fracture, sublimation, and replication. (*J. Boumendil, CMEABG UCB-Lyon 1*)

3.5.3 Comparison of "Chemical Fixation," "Physical Fixation," and "Cryo-embedding" Techniques

Chemical Fixation ("Techniques" Chapter 2, Section 11), "Cryofixation" Physical Fixation ("Techniques" Chapter 2, Section 12), and Cryo-embedding ("Techniques" Chapter 2, Section 10)

Bulk material: collagen fibers in a tissue (Figs. 8.56 and 8.57)

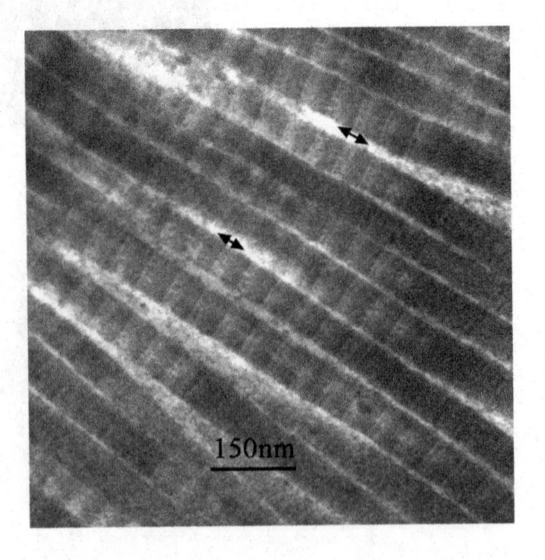

Fig. 8.56 Collagen fibers in a tissue after chemical fixation, embedding, ultramicrotomy, and positive-staining contrast. (*J. Boumendil, CMEABG UCBLyon I*)

Fig. 8.57 Collagen fibers in a tissue following high-pressure cryofixation and cryosubstitution in the presence of osmium, embedding in Lowicryl at low temperature, ultramicrotomy, and positive-staining contrast. (*J. Boumendil, CMEABG UCBLyon I*)

Comparison discussion: These are type I fibrillary glycoproteins that are found in a bulk tissue where epithelial tissue and connective tissue are often found together. This type 1 collagen is present in the form of large fibers, characterized by a periodicity of 64 nm, which corresponds to its chemical composition and its spatial conformation, which is clearly visible in both examples. This periodicity is highlighted by the reaction with the osmium. After chemical fixation (Fig. 8.56) and embedding in epoxy resin, the fibers are better individualized than after cryofixation (Fig. 8.57) because in the first case, the other types of associated collagen and other non-fibrillary glycoproteins of the connective tissue are preserved less well. Conversely, immunolabeling will be more intense in the case of cryofixation, since more antigenic sites are preserved.

What is characteristic in collagen is the periodicity. This periodicity serves to validate the classical technique of chemical fixation, embedding, and ultramicrotomy at the beginning of microscopy, by comparing the known results using X-ray diffraction.

With the six techniques applied to collagen, chemical fixation, cryofixation (followed by embedding and ultramicrotomy), freeze fracture, decoration shadowing, negative staining, and immunolabeling, we have a set of tools for viewing, recognizing, and localizing a protein whether it is in the form of fine particles or bulk material. These six techniques are complementary.

Negative staining yields a result very quickly. Shadowing highlights the smallest structures. Immunolabeling will be carried out directly on isolated fibers viewed using the negative-staining technique. It can also be used on a section when it is a bulk tissue and will be more intense on sections obtained using cryo-methods (PLT, cryo-ultramicrotomy). The direct observation of collagen fibers in a frozen hydrated film (no example available) will probably allow for more detailed results on the structure of different types of collagen.

Bibliography

Abid, M., Ehret, G., and Touroude, R. (2001). *Appl. Catal. A*, 217–219.

Abid, M., Paul-Boncour, V., and Touroude, R. (2006). *Appl. Catal. A*, **297**, 48.

Ayache, J. and Albarède, P.H. (1994). *ICEM 13*, vol. 1. Les Editions de Physique, Paris, 1023–1024.

Ayache, J. and Albarède, P.H. (1995). *Ultramicroscopy*, **60**, 195–206.

Le Cam, E. and Delain, E. (1995). Nucleic acid–ligands interactions. In *Visualization of Nucleic Acids*, vol. 18 (ed. G. Morel). CRC Press, Boca Raton, London, Tokyo, 331.

Chapter 9
Conclusion: What Is a Good Sample?

The act of preparing each sample, as well as observing it in the TEM, subjects the material to various stresses. Each sample contains both intrinsic defects, giving the material its particular properties to be studied, and extrinsic defects, which are associated with the different stages in its history (i.e., from its preparation to its observation in the TEM). After passing under the microscope, each sample will have undergone modifications that must be recognizable so as not to bias the interpretation of the structural and chemical information obtained.

The important thing to be aware is that the sample being analyzed never corresponds to the original material we are investigating. Artifacts, whether visible or not, are undesirable defects and are always present in a sample. Their formation cannot be prevented because they are inherent to a technique; they can only be minimized. What is observed in the TEM is an infinitesimal representation of the sample, if the surface area and the volume analyzed are taken into account. The process of turning a material into a thin slice already gives the material a history: It will have undergone a certain number of stresses (mechanical, chemical, ionic, physical) that may have modified or transformed it. Another part of the material's history takes place during its observation in the microscope under conditions that will also add stresses. In the best-case scenario, microscope observation will not damage the material; however, in other cases, serious damage will occur. Depending on the type of material, this damage can result in the destruction of the sample under the beam. Despite all of this, we are still able to observe and analyze the sample in the end.

Before returning to the microstructure of the initial material and the consequences of its behavior, properties, structure, etc., it will be necessary to use a combination of several preparation techniques to verify the validity of the structural, chemical, or spectroscopic information retained in order to describe the sample.

In materials science, a good sample, i.e., a thin slice of the material, is not a perfect sample. This means that it is not without defects. It is necessary to know and recognize the intrinsic defects of a structure and, consequently, to be able to select a preparation technique consistent with the type of analysis to be performed.

Artifacts are more numerous in biological materials than in materials science. Chemical fixation, the step most often used for preparing any biological material,

J. Ayache et al., *Sample Preparation Handbook for Transmission Electron Microscopy*,
DOI 10.1007/978-0-387-98182-6_9,

immediately introduces artifacts that modify the initial structure. Knowing the effects of each chemical component used on the structure is fundamental to good interpretation.

In addition to all of this, the material's response to the application of the technique and the artifacts created takes the chemical, physical, mechanical, etc., properties of the material into account. They can be used as indices for understanding the problems related to the structure–property–function relationship.

In TEM, the sample preparation technique is always cumbersome and stressful for the material. At the moment of observation, our critical mind is of paramount importance because we know the perfect sample does not exist. The microscopist therefore must

- completely master the technique in order to prevent the introduction of too many modifications to the sample;
- choose the best technique with regard to the material and the problem posed;
- recognize the unavoidable artifacts of each technique in order to deduct them from the interpretations; and
- during TEM observation, select the thin slice areas that contain no artifacts or as few artifacts as possible.

Photo Credits

Abid M., LMPC-ECPM-ULP Strasbourg, FR
Ayache J., CNRS-UMR 8126, Villejuif, FR
Baconnais S., CNRS-UMR 8126, Villejuif, FR
Badaut D., EOST, Strasbourg, FR
Benaïssa M., IPCMS, Strasbourg, FR
Boumendil J., Université Claude Bernard-Lyon 1, Villeurbanne, FR
Buffat P.A., EPFL-CIME, Lausanne, CH
Burdin B., CTμ UCB-Lyonl, FR
Cantoni M., EPFL-CIME, Lausanne, CH
Cattan E., Université de Valenciennes, FR
Carl Zeiss NTS, Oberkochen, GE
Collin S., CSNSM-CNRS-IN2P3, Orsay, FR
Cojocaru C.S., IPCMS, Strasbourg, FR
Cosandey F., Rutgers University, Piscataway, USA
de Chambrier S., EPFL-LESO, Lausanne, CH
Delain E., CNRS-UMR8126, Villejuif, FR
Dieker C., EPFL-CIME, Lausanne, CH
Ehret G., IPCMS, Strasbourg, FR
Fisher Bioblock Scientific, USA
Ganière J-D., EPFL–IPEQ, Lausanne, CH
Gnaegi H., Diatome, Bienne, CH
Haftek M., EA 3732 – UCB-Lyon1, FR
Jéol., Ltd., Tokyo, JA
Laub D., EPFL–CIME, Lausanne, CH
Iliescu M., IPCMS, Strasbourg, FR
Lattaud K., IPCMS, Strasbourg, FR
Le Cam E., CNRS-UMR8126, Villejuif, FR
Longin C., INRA Jouy-en-Josas, FR
Michel N., LEMHE, Orsay, FR
Pehau-Arnaudet G., Institut Pasteur-CNRS, Paris, FR
Qiang F., Fritz Haber Institut der Max Planck Gesellschaft, Berlin, DE
Radmilovic V., NCEM, Berkeley, USA
Rivoire A., EZUS UCB-Lyon l, FR

J. Ayache et al., *Sample Preparation Handbook for Transmission Electron Microscopy*,
DOI 10.1007/978-0-387-98182-6, © Springer Science+Business Media, LLC 2010

Ryter A., Institut Pasteur Paris, FR
Schüler A., EPFL-LESO-PB, Lausanne, CH
Sizaret P.Y., Université de Tours, FR
Song Cheng Y., NCEM, Berkeley, USA
Touroude R., LMPC-ECPM-ULP Strasbourg, FR
Werckmann J., IPCMS, Strasbourg, FR

Index

F

P